Are We There Yet?

Diary of a Project Manager

The Good, the Bad and the Utterly Random
Concerning One of the Most Essential,
Unpredictable and Unsung Responsibilities in Business

by

Donald A. Pillittere

Second Edition

Are We There Yet, Diary of a Project Manager

Donald A. Pillittere

Copyright © 2013 by Donald A. Pillittere

ISBN: 978-0-9851942-3-9

Paperback
Published by Donald A. Pillittere
6 Bella Via Lane
Spencerport, NY 14559
USA
dpillit1@rochester.rr.com

Printed and bound in the USA

Cover design and book layout by Mary Bartholomew

To my mom Joan Pillittere who loved me no matter what I did or said to embarrass her, which I did all the time. I dedicate this to her as well because she loved my wife and children even more.

To my father-in-law, Edward Annechino, who gave of himself without question to his wife, children, grandchildren, family, and friends.

Thanks to both for setting examples of unselfish love which I try to emulate each and every day.

To my family: my wife, Lori, and children, Donald, David, and Julie, you make my life sunny.

Are We There Yet, Diary of a Project Manager

Table of Contents

Introduction

My hair has turned prematurely gray. When friends ask why, I put down my oversized bottle of industrial-strength antacid tablets and explain that, for much of my professional life, I've worked as a project manager. "What's that?" they ask, and I begin my answer. Soon after, they pick up the bottle, stuff a handful of tablets in their mouth, and suddenly remember an important thing they have to go and take care of.

Everybody has been a part of a project at some time in his or her life. Your entry in the junior high school science fair was a project. Dinner for twelve is a project. So is mowing the lawn, creating a scrapbook, planning a vacation, building a house, pursuing a cure for cancer, raising kids, or developing and delivering a new piece of technology.

With any project, there are several common requirements:

- A goal or objective
- People to do the work
- Money to pay for everything
- Materials to give the project shape
- And planned tasks to move the project along to completion

Projects have all these things and one other thing as well. All projects have problems. There simply is no such thing as a perfect, problem-free project. If there was such a thing, many more project managers – including me – would be dark-haired, unemployed and spending a lot less time in the drug store antacids aisle.

When all is said and done, most of these problems are people-related. Think about how hard it is to get two people to agree (if you need evidence, look at the divorce statistics), then imagine how much more difficult it is on a project team – which is what we're concerned with here – comprised of ten, twenty or more members. Team members come from different "cultures," rely on their knowledge of particular functions, and like to stay within their comfort zones – often at the expense of the "big picture." It's the most natural and the most frustrating of human behaviors.

This tendency can challenge the best project leader or manager. Every well-conceived project has a designated employee who is responsible for getting team members to "play nice," keeping the project goal in sight, and making everyone part of the solution. The project manager's task is especially daunting if there are no formal rules or procedures that help guide the team on its journey. And that's more common than you might think. In many cases, the project – a product based on new technology, for example – takes the team into uncharted waters. So the rules and processes are often written "on the fly."

Even so, I actually enjoy working on projects. It's taken me almost twenty years to realize that. But why not? For me – indeed, for a lot of people – there's nothing better than setting a long-term objective that will stretch and enhance your abilities, and then actually achieving your objective and making something better. Even with all the frustrations that come with a project, there are some terrific benefits – like meeting new people, learning new skills and accomplishing something worthwhile.

Each project provides a unique experience. Each has its own particular goal, working culture, internal processes (or lack thereof), people on the team, and ability to handle the inevitable crises that are an all-too-common impediment to progress. For that reason, it would be difficult to come up with a specific set of "silver bullet" rules or procedures to assure the success of every project – although there are some good "rules of thumb" that can be applied to virtually any project.

My goal for this project – this book – is to help you the reader, appreciate the benefits of project management, prepare for problems that are certain to arise, recognize some of the factors that are consistent among projects, and keep a sunny outlook even (especially!) when the thunderclouds are looming. Oh, and one more thing: I'd like you to etch "The 4P's of Project Management" into your brain. Marketing has its "4P's": product, price, place and promotion. So I've decided to create 4P's for project management. They are:

- **Process** – your project's "rules of the road".
- **People** – by far your greatest asset, and probably your greatest headache.
- **Parts** – all resources (animate and inanimate) available to your project.
- **Phenomena** – events, issues, attitudes and who knows what else that will work to propel or stall your project, think Murphy (and his Laws) after using Barry Bond's clear performance enhancing supplements.

These are the key elements that must be addressed – indeed, they can hardly be avoided – if you're going to have any chance at a successful project, especially one that you're managing. Project managers are given responsibilities along a straight line – that is, you have to get from where you are to where you need to be. And, along with your own skills, ingenuity and grace under pressure, it's the "4P's" that are going to get you there.

Your boss is on the left, your administrative assistant on the right, and a project team arrayed around you. Everyone approaches this job in their own particular way and within the guidelines they receive from the organization (and its intrinsic culture). Organizations that provide the process, people and parts, as well as the means to address phenomena, give their project managers a running head start. After that, it's up to him or her to lead the team to their goal – a new product, service, or some other achievement that enhances the revenue and stature of the company.

Keep this in mind. Think about what's presented here in the context of these essential factors. View this book through the prism of the "4P's of Project Management," and you can't help but come out of this experience far more knowledgeable and better equipped to succeed at a project of your own.

What follows is a project manager's diary – the day-by-day chronicle of a project team developing a series of new products. This account is based on true events that I witnessed and participated in over the years at several different companies. The names of the places and the people involved have been disguised to protect the guilty (as well as my own future employability). But the essence of the experience, the challenges and nuances of project management, are otherwise as real as I can make them.

So here's a tale about what can happen when talented people with good intentions set off on a great adventure – with no map and Murphy's Law lurking around every corner. All I can say is: That's project management. It brings a whole new meaning to that familiar old saying: "Are We There Yet?"

Chapter 1

"I'm new to the team and very optimistic. I lack product/peer knowledge. And I want to fit in, so I go along with the best case scenario for the project. Does that make me a problem? Hey, I should fit right in."

June 2

Dear Diary,

Well, I got the job with the communication company. Meet the Manufacturing Project Manager for a new portfolio of products that will incorporate the latest communication chip set from a Fortune 100 company. I'll oversee two distinct product families: one for low-frequency bands and the other for high-frequencies. Basically, the products are a handful of circuit boards, enclosures and software. The majority of the mechanical parts will be carried over from the previous generation of products. Only circuit boards and software are part of the commercialization effort.

I have a lot of catching up to do, since the program has been in progress for well over six months. The low-frequency program is further ahead, and Operations – which includes manufacturing, testing, purchasing, logistics, and me – has to start planning for production. I'm excited to get started on this new adventure, and maybe a little apprehensive. "Newbie" to team or not, I already can see signs it will be more like a roller coaster ride, and I despise roller coasters. I've been given a new version of Microsoft Project – which I haven't used in more than ten years – to develop a schedule. I need to become an expert real fast.

It occurred to me, after a handful of meetings during the first week, that Sales has sold products that don't exist yet, and as far as I can tell, won't be ready in time for their promised delivery dates. In fact, the Design Manager, Dean Sign, made a bold and not-too-encouraging statement during one meeting that the new products won't be ready until late next year (I'm interested to see the accuracy of his prediction). Bolder yet, no one seemed to pay much attention.

But that's not new. All projects, even presented with bad news, have their share of unbridled enthusiasm – most often from Sales, but even from people who should know better, who've been on numerous projects throughout their careers (like me!). We never seem to learn. Maybe it's the high from being on a new project that clouds the realities facing the team. Maybe we say what the boss wants to hear. Maybe it's easier to think happy thoughts than to anticipate the next eighteen months filled with delay, blank stares and pain. Where did I put my oversized bottle of industrial-strength antacid tablets?

June 6

Dear Diary,

Our VP of Sales, Sal Moore, really knows the market and is pushing us to succeed. That's good and bad. Sal seems to have overcommitted the team regarding availability of low-band products, and it will be up to me to see how far off his commitments are once I complete a project schedule. But after seeing Sal in action, I'm glad he's on our side. As much as everyone loves to hate Sales, they're the ones bringing in the dollars that pay salaries and benefits. Sal has personal relationships with many large customers and he's taking full advantage of the situation.

I got to work gathering information on the boards and put together a schedule that showed critical paths and, ultimately, when pre-production units could be built. Each board requires a schematic, bare board layout (and files), a bill of material (BOM), parts specifications and an assembly procedure. It isn't brain surgery, but things can go wrong since lead times for parts are unknown in the beginning. Bare boards can be purchased in three days or less (at a premium … ouch!), so there are no big worries around board lead times – just the parts. The good news is that each of these boards is on its second or third revision, so lead times on existing parts are known. The bad news is that Engineering has been known to change parts on, or add new parts to, the BOM without much notice.

I had a chance to question the design engineers and production team, and the structure of the low-band products has become much clearer. So I've created a "cheat sheet" in Excel with the board name, the design engineer responsible and the corresponding upper-level part number. I also put together an Operations schedule, and if all goes well – one happy soul on the team put it this way: "If we recreate the 1980 US Olympic Hockey Team's 'Miracle on Ice'" – we could meet some of Sal's dates. I'm skeptical because we haven't done any "what if" analyses. Based on my experience, even the best-run projects rarely launch on time.

Good thing, I suppose, that our competitors are operating under essentially the same challenges and dynamics. Of course, that doesn't change the fact we're in a race against them for customers, revenue, etc. As Sal likes to say: "Losing is not an option."

June 11

Dear Diary,

What a surprise … The dates provided by Engineering were optimistic – in the extreme. Not that there was any misrepresentation intended. They just didn't put a lot of thought into potential "show stoppers." Target dates are great goals to shoot for, and nothing would be better than achieving the impossible (think of the bonus!). But durations for a lot of our tasks – approval of bare board layouts, parts lead times, build times, etc. – weren't properly padded for "what if" situations. Case in point: What if Design is a week late in approving new board layout? What if there's a newly specified part with a long lead time? What if the supplier doesn't have the capacity to build boards as quickly as we need them? I think I'm developing a rash.

Being new to the team isn't helping matters. My cheerleader optimism is going to have to give way to a bit more realism. My lack of specific product and people knowledge can be overcome only with time. On top of that, my experience tells me to plan for the unexpected by adding time to certain critical tasks, but my human side wants me to fit in with the team and go along with the best-case scenario. So far, the best-case scenario is winning.

Meanwhile, as if we don't have enough issues, our project team is scattered over four different time zones on two continents. Multiple locations adds a hobgoblin or two to our festering little nightmare: If one location has a normal number of issues, two locations doubles those issues (2x1 = 2), three locations expands that number (3x2x1 = 6) and so on. It is one thing to get a due date from someone you're dealing with face to face and quite another if the person is half way around the world. Besides, I don't know who these guys are, how they work, or their strengths and

4

weaknesses. Can I trust their estimates to get critical tasks completed?

Even so, my job at this point has been relatively easy since it's primarily just collecting what others give me. The engineers have a much more difficult task – trying to figure out how soon a very complicated design can be finalized.

June 16

Dear Diary,

I debated over the weekend whether or not to modify the schedule and add a few more days to critical tasks, like approval of bare board designs and parts lead times. I guess the realist hasn't beat out the cheerleader just yet – I decided against it. In one of my earlier jobs as a manufacturing engineer, when it came to estimating product costs, sandbagging the numbers to guard against a sharp change in volume was a necessary evil. Most of the Operations people never really trusted the Sales forecast anyway. But for this project, we agreed to review the schedule weekly, so delays of any kind should be obvious (shouldn't they?).

Hopefully, it will be a simple matter of how soon the team can overcome delays and get back on track. This schedule is going to be shared with all of the managers and the CEO. Still, it's tough going against my gut feeling that things are just not going to follow the schedule. Only time will tell whether the grumpy realist or the cheeky cheerleader is the better prognosticator.

Meanwhile, it's only my third week on the job and I traveled to the West Coast. I got the opportunity to travel with our supply chain guru, Advanced Supply Chain Manager, Peter Parts, to meet with our contract manufacturer (CM). None of our company's sister divisions has available production capacity, so the decision was made months prior to outsource production. Fortunately, months before I arrived, an Operations team traveled worldwide to assess contract manufacturers, and this company was the winner. After what I'm told was careful consideration, they were selected and approved by our Business Unit Director, Richard Cranium.

A two-day tour of the company left me very impressed with their capabilities. They can perform design for manufacturing (DFM) and design for test (DFT), develop test fixtures, procure parts, manage new product introductions (NPI) and transition product to China to take advantage of low-cost labor. We'll need all of the above if we're going to have any chance of building, testing and shipping a quality product relatively quickly.

I really am excited to be working with this CM. As remarkable as their capabilities are, their people are even more impressive. In my experience, the success of any project often hinges on how well people – both internal and external to the team – get along, and this initial interaction with the CM tells me we have a good chance for success. I even have some hope that handing over some of the tasks to the CM will compensate for my concerns over the lack of "what ifs" built into the schedule, and my not adding in any slack time.

June 20

Dear Diary,

We're still developing our BOMs, and the final product structure in our MRP system is sparse at best. So we have to order parts "off system" – which means going to Richard for approval to purchase parts. This doesn't make sense to me, since we have Operations VP Manny Factura. When I became aware of this requirement at a team meeting, I added a task – "parts procurement approval" – against Richard's name. How cool is that, delegating upward!

No one in Operations has to look at the schedule to know the importance of ordering parts for low-band boards. We've all been part of previous commercialization efforts and understand the benefit of getting a jump-start on the supply chain. We all know a project schedule ought to have a parallel supply chain schedule. It's one thing to know product design will be completed in six months, but this information doesn't do much good if the lead time for a critical part is another six months. If the supply chain isn't managed in parallel with the project schedule, then nothing will reach the customer for a year. But if lead times are known early, parts can be ordered and arrive "just in time" for pre-production – essentially giving us a head start on production and competition.

Peter Parts has been aggressively quoting new parts recently, getting lead times and determining the cost to purchase initial production quantities. He was smart enough to push through a purchase order for a new integrated circuit (IC) that will be the heart and soul of both the low and high-band products. This could be risky if our project's direction changes or the company decides to redesign the IC. If not, it's a great move. Besides, to have a chance of staying on track, we need these parts sooner rather than later.

Peter created an Excel spreadsheet, to be shared with Richard, that tracks our current board costs and estimates labor requirements as well as total units required by Design and our key customers. He also covered some of the longer lead-time parts. Peter and I are simpatico on a lot of things and quickly formed a partnership. His drive to get lead times helps make my schedule dates more accurate. As a Project Manager, it's great to have supplier quotes that back up durations for some of the important supply chain tasks.

At this stage of the project, Peter and I scheduled a meeting with Richard – actually it was a video-conference – to get his approval to purchase parts for low-band boards. This was my first interaction with Richard since my interview almost two months ago. I really had no idea what to expect. He immediately let us know that he wasn't happy with the cost of buying parts. A discussion ensued about whether Peter double-counted parts we already had on hand. It was a valid question, but I don't think we can afford to delay a decision while Peter goes back and readdresses his data.

To try and keep things moving, Peter mentioned that costs would go down once volume purchases can be made – but to no avail. No one has ever accused me of excess patience, and I know what the schedule shows. We needed Richard's approval right away to move forward and purchase the initial parts. I was getting a brain cramp thinking about how Sal had committed us to unrealistic customer due dates while Richard was delaying a "buy" decision to get more details on board cost. How can two people in the same company be making contradictory decisions so early in the project?

It wasn't the first time I'd run up against this conundrum in my Project Manager career. But that didn't make it easier to deal with. Other companies are competing against us as we try to get product to market, and now Richard (of all people) is holding us back. At the end of the video-conference, we agreed to look into the numbers one more time and meet again next week. Of all the things I thought needed some added slack time – final board layout, for example – getting approval from Richard wasn't even a blip on my radar.

To be fair, everyone is trying to do their best to commercialize our products. But it feels like we're helpless right now against some of these arbitrary decisions, so we struggle. Trying to follow Richard while he apparently is still sorting things out in his own head is burning through our budget and creating a lot of frustration. Meanwhile, Sal is making promises about new products to his customers that haven't been communicated to the project team. With e-mail, cell phones, Blackberries, fax machines and the rest, you'd think Richard and Sal would get together now and then. But you'd think wrong.

June 23

Dear Diary,

The Production Management Team – which is Manny, Peter and I – wanted to meet with Richard before the weekly Wednesday project team meeting to share new data on parts lead times and our ability to get boards built. We set the meeting up for Tuesday afternoon, hoping for a "quickie" and confident it would be a "slam dunk", thanks to our updated numbers. So we're less than five minutes into the meeting and Richard picked up right where he left off – asking for more detail on expenditures. Early on, it became clear to us that Richard wasn't going to give us anything remotely related to a "go ahead" that day.

Then, to make absolutely certain our heads exploded; Richard told us he was pressuring Sales to generate more revenue with our new product. So now he's pushing both ends of the business all on his own – applying pressure to simultaneously drive down the cost of and boost the revenue from a non-existent product, which at this point is little more than a concept that exists only in certain people's heads, since Marketing and Design rarely talk directly to each other.

Richard is Bottleneck #1. True, each time he steps in we save money, but he's going to save us to death. What potential long-term revenue is being lost by his involvement? Who knows? Project Management 101 teaches that all business decisions should be based on cost/benefit analyses. In this case we know the projected cost savings, but we don't know the lost revenue.

We pleaded with Richard. We even sent him my project schedule with his name and the various tasks his decision would affect – highlighted in yellow. Lead times for some parts are 16 weeks, and chances of meeting schedule dates are falling swiftly. And here's the thing: The problem is us! Vast amounts of delay caused by internal problems – poor planning, inadequate communication, mismatching skills to tasks, leadership (that is, the lack of it) and poor understanding of customer needs. Richard was micromanaging the team big time and it was hurting our chance of getting working product to market on schedule.

Addendum, June 23

Dear Diary,

Sorry, I'm still stewing about these past few days. I think it's safe to say the honeymoon is over. What's become very clear to me at this point is that our team doesn't have a well-defined goal. There are no formal product family plans, no marketing documents and no business case. I've seen some drafts of product requirements from Design, but they're mostly industry-standard boilerplate. A lot of what we say about and expect from these new boards is intuitive or assumed. I'd wager that if you asked three different team members what we're trying to accomplish and why, you'd get three different answers.

I'm more used to the well-defined commercialization process we had when I worked at a Fortune 100 company. At times it drove me crazy, but I'm starting to miss "the good old days." Those processes occasionally led to "analysis paralysis" – spending too much time trying to get something perfect and missing a market window (sometimes "good enough" really is good enough). On the other hand, our processes lack pretty much anything related to practical analysis and formal structure. My schedule is an attempt to get everyone thinking about some well-defined end date. But everybody is too wrapped up in their own worlds – their own perspectives and priorities – to see the impact they're having on time-to-market.

We have some really smart people here who know the market and tend to work in an ad hoc fashion – which is fine for teams working on an innovative product like an Apple - **iPod**, who get all the time they need to get the job done. But our situation is different. Customers are expecting products based on industry standards, and a couple of our competitors have already issued press releases articulating their plans – with product availability dates. Everyone knows what's expected based on the standard. The only unknown is which company is going to get to market first. As far as our company is concerned, that's still a mystery.

We're like an orchestra with members all playing different pieces of music, and the end result sounds dreadful! At least Operations is trying to get everyone to play the same tune. After that maybe we can improve on our sound and timing. Being an engineering-centric company, Operations isn't viewed all that

11

favorably, even though our team does an excellent job. In essence, we're the catch-all that's expected to compensate for problems with Design, Marketing and Sales. Management is constantly squeezing the supply chain bubble, assuming it won't burst. That's not to say we're perfect. We definitely have our share of problems. All functions, even Operations, has room for improvement. It's just one of those facts of life that the buck tends to stop here.

OK, I feel better now … not really. I guess that'll do it for the cheerleader in me. Enter the realist, and it hasn't even been a month.

June 29

Dear Diary,

That ominous, relentless sound you hear in the background is a ticking clock. Even though Richard promised to give us approval to purchase parts, the decision has yet to be made. At Wednesday's team meeting I bring up the fact that we have now lost two weeks of lead time, and no one – not even the managers – is willing to give Peter the go-ahead. It's a good thing Peter is experienced. He's already instructed the buyers to go ahead and order longer lead-time items. His motto is: "It's easier to get forgiveness than permission."

We still have a long way to go. My optimistic outlook on schedule dates is proving to be very wrong. Our engineers have missed due dates on two of the four boards, and recent changes that required new parts have made availability dates for them a crap shoot. Dates for tasks and durations are changing as rapidly as the board designs. Engineering change orders (ECOs) are flying fast and furious since Operations made it clear to Engineering that we were going to use the BOMs in our MRP system as a purchasing guideline. But there's nothing easier than changing a change order, so who knows?

I like the project schedule because it's a relatively simple way to at least try and drive some discipline. Training the engineers to use the formal ECO process to update BOMs is another. Released BOMs are needed to track change histories, adhere to ISO 9000 standards and store drawing files. But just as important, it ensures that we're buying the correct parts based on the right specs. If and when we transition to a CM, these documents are going to be the "recipe" for the product, and you can't "add salt" to improve the product later on. We have to guarantee accurate data.

This Week's Sign That the Apocalypse is Upon Us: During the project meeting, the engineers ask when they're going to get boards to test. Say what!?! The schedule shows that half of the boards don't have approval from Design, there are new parts being quoted for lead time and cost, and good ol' Richard still hasn't approved initial purchase. I asked if anyone has actually read the schedule and was met with dead silence and blank stares. What's the use of a schedule if no one is reading it? I asked if there's anything I can do to make it easier for Engineering to keep track of schedule and sharing dates. More silence.

I tried one more thing. "What if I take only our critical path items", I said, "and generate a secondary schedule to minimize the current complexity and get us more focused on immediate needs?" This last attempt got buy-in from Engineering. Maybe it'll help. In addition, I've created a separate spreadsheet with ongoing action items for us to review daily to help keep us on track. Even though the document is viewed by the team mainly as a list of Operations action items, I've pulled all the functions into the mix.

Somewhere along the way I learned that, sometimes, a project manager has to be, well, sneaky to get what they want — like grinding up and hiding vitamins in a kid's pudding.

As frustrating as the project has been, I suppose none of us who've spent any amount of time in manufacturing and supply chain should be surprised. Maybe we're gluttons for punishment. It's clear to me now, that no one is formally driving this project toward a common end goal. Operations can certainly play a role, especially if we can get the larger team to follow the critical path schedule and find enough proverbial fingers to plug all the holes in the proverbial dike.

But there doesn't seem to be a sense of urgency within the team, even as Sales beats us up about delivery dates. I suggested to Sal that he attend our next team meeting to get a handle on progress. He basically ignored me and probably will continue to call and e-mail me, as if that's going to get him product sooner. I call this Sal's "Field of Dreams": Instead of "Build it and they will come," Sal says: "E-mail 'em with due date and they will build it."

Right. Problems like these are continuing to delay the schedule, and Operations is continuing to analyze ways to save time to keep the project on track. Gluttons for punishment? Maybe. But, hey, this is what I do. In the end, there's nothing better than seeing the product finally ship out of the factory, knowing all the hard work the team has put into fulfilling customer needs. I think maybe we just had a bad month, and things will get better in July. (The cheerleader returns …)

Learning #1: Process

Begin at the End

Every project has to have a clear end goal or objective. The success of a project largely depends on how well the goal is defined at the beginning. This might be the most critical step, and it requires accurate data to help define the end result. Without a clear end goal, the project probably will fail, or at best be minimally successful. It's at this early stage that the battles among team members must be fought and resolved. There's no better time to get a realistic assessment of the project and its potential for success. The more time spent at this stage understanding customer needs, scanning the competitive landscape, uncovering potential showstoppers, and determining the time and cost of the project, the higher the probability that your project will be a winner.

Tabrizi and Walleigh studied twenty-eight next-generation product-development projects at fourteen high-tech companies. They found only four companies were successful in meeting their upfront expectations. Those that were successful put tremendous effort into developing a clear product map – covering markets, specifications, budgets, organizational issues and technology requirements. Maps were developed without any holes that competitors could exploit and with thorough market and technical data. For the companies that failed, their downfall could be traced to shortcomings in this initial definition phase.

Another key issue: Even though all the companies in the study had product maps, the four successful ones treated their maps as "living documents" and referred to them regularly to ensure active communication, commitment and focus throughout the organization. Everyone knew the end goal and the plan to get there, including potential roadblocks to success. Contrast this to the product map that was in the mind of Richard, Sal and Dean — each had their own, and each was very different. No wonder their team is so lost. Crazier still, is how each of these guys keeps looking to Operations for answers on availability of a product that hadn't even been designed yet.

Each assumes they know and are following the plan, and that the other guy knows and is following the plan as well. But as is becoming clear, all the key managers have their own version of the plan in their heads. This leaves the project team wandering somewhere in the ether. In this case, the team needs a litmus test to help it make decisions. It needs basic planning tools, such as a product portfolio, projected product volume and lifecycle, selling prices, target unit manufacturing costs, and a capital budget so Operations can recommend the best possible actions.

Our happy little team doesn't seem to have a head start on much of anything. Thank the project management gods, this isn't where my story ends.

Chapter 2

"Project management is like the popular arcade game Whac-A-Mole, where the moles are the critical items that need to be knocked down or eliminated before the next critical item rears its ugly head. Let's just say that, right now, my score isn't in the Top 10."

July 2

Dear Diary,

At first glance, things are looking up (cheerleader!). But something tells me this is just the relative calm before the storm (realist). Design is off doing its best. Manufacturing is ordering parts after – finally! – receiving Richard's approval, three weeks past the scheduled date. Changes are coming pretty fast. In fact, I've been having a hard time keeping up with all the numerous schedule updates required to show everyone what the current critical path looks like.

After the cost beating he took from Richard and the three-week delay getting approval, Peter licked his wounds and started pushing for projected sales volume – and promptly ran into a brick wall. Even though Sal has been off selling, he doesn't want to commit to a high-level forecast. He just uses that old sales adage – "These customers are very critical and worth hundreds of thousands in sales. We must ship product immediately!" – to push for trade trial units. This might be what they teach in Sales 101, but it doesn't generate the kind of volume that will drive parts, labor and equipment costs, as well as capacity requirements for production.

I'm starting to rethink my high opinion of Sal, since not having volume is going to cripple our long-term operational tasks and decisions – one of which will be addressing the board and product costs with Richard. So once again, our managers are looking at different end goals and Operations is going to be caught in the middle - Yippee.

At this time our costs are nowhere near the targets projected by Design. (Yes, Design projected cost targets. Go figure.) We're continuing to limp along, hoping for a discount on parts, even though we still don't know volume. Everyone is so concerned with getting product to the initial customers, no one has figured out whether this dog will even hunt – especially when it comes to market price, standard channel discounts and margin. The lack of well defined volume and price targets (both UMC and sales price) is simply unbelievable.

In a former life as a Worldwide Product Manager, I lived by the business case that included volume, sales price, UMC, R&D, SGA, cost, etc. They were reviewed at every gate, and any change that caused the Net Present Value (NPV) to go negative could lead to a decision to cancel the program. I'm thinking about developing a business case here, but I'm too busy trying to keep my head above the project chaos. Besides, I'm not sure how to input imaginary numbers into Excel. Would anyone look at it anyway?

Design has made some headway with the products, which actually appear to be working! Our working definition of "working," of course, falls well short of industry specifications "working" and – more importantly – customer expectations "working." On the bright side, our team has a chance to be optimistic, if only for a little while, since some progress is actually being made. When and whether the product will actually work in a more robust (that is, more generally acceptable) manner, is another story. So far, no one wants to take a guess at when that will happen, even though Richard asks us weekly.

Yesterday Richard commended the team on its progress, even though everyone knows we're way behind schedule. It's not just in incorporating industry-standard features, but – just as important – meeting safety and EMI regulations. I found out that one of our sister companies is putting together a project to address RoHS/WEEE directives. Operations mentioned this new directive at the last team meeting to see if it's something we need to address. True to form, we were ignored.

July 8

Dear Diary,

Talk about adding insult to injury. Not only are we way behind schedule on the low-band product, but RoHS/WEEE directives are looming on the horizon. The team wasn't even fazed by Operations' questions about whether our products had to be certified. Meanwhile, Peter and I joined a team at one of our sister companies to address this issue in parallel. "Sister" has a mountain of resources to tackle this, and we're going to need all their help to resolve the issue. During the week we also attended a handful of meetings to better understand the RoHS/WEEE directives and brainstorm ways to tackle them in parallel with the low-band products.

Our weekly team meetings are starting to remind me of the movie *"Ground Hog Day"* starring Bill Murray. We keep waking up with product in the same situation it was the previous week. Richard must love this flick! He makes for a very believable co-star. Every week he asks for some documentation around milestones – a schedule from Design, perhaps – and nothing comes of it, and he's just as satisfied to hear that the product is working, whatever that means. And the next week he asks again, and … Meanwhile, from my perch, I see little discernable progress.

My request to my boss to be appointed overall project manager has gone nowhere. So I sent an e-mail to Dean, the head of Design, offering my help after a recent team meeting where Richard asked him about … engineering milestones! Timing is everything. It'll be interesting to see if Dean takes me up on my offer.

July 13

Dear Diary,

We've been at this for less than two months, but even so, I wondered how long it would take. The Great Savior of all corporations imprisoned by myriad "process" problems – namely, a reorganization! – has been initiated. This will fix things. Or, to borrow words from actor Mike Myers, "Not!"

The gang in Manufacturing has been moving forward with parts procurement, paying no attention to this reorganization. No surprise there. A lot of us have learned to ignore these types of activities and focus on project deliverables. We've been trying to get parts ahead of schedule to make up for delays caused by Richard and to compensate for newly specified parts from Design. Trying to understand – or even think about – reorganization at this point would delay us even more. There is an urgency for Operations to get ahead of the curve (and force the bottleneck to somewhere else in the program). It would certainly help our standing in the organization and maybe garner some respect from Richard, Dean and Sal.

And speaking of change: Richard has decided that we're going to use a different contract manufacturer – ACME Inc. – instead of the excellent California outfit that was originally selected. Why? He wasn't exactly clear on that point. I'm not happy. It's always hard to foster a relationship with a supplier – but then pull the rug out from under them? Not wise. But I've seen it happen a number of times before. So my partner Peter and I have more work on our plate, putting together a document package so ACME can quote. No big deal, actually. With today's technology, a series of e-mails with attachments gets the job done in a couple of hours. We have no idea what to expect from ACME, but we do as we're told. I hope Richard knows.

July 19

Dear Diary,

This week, we approach two milestones (and hope we don't trip over them). We have to deliver working products to 1) Design and 2) one of Sal's customers. The schedule shows that, to even have a chance, we need final BOMs for a majority of boards. Peter rolled the dice and ordered the old-revision boards with the hopes of getting new changes as they happen so he can place purchase orders. Umm, looks like "snake eyes."

As much as I hate to deliver bad news, I told Sal it will be impossible to get a unit out to a customer by the end of the month. He wasn't hearing any of it – even after I sent him a portion of the schedule in Excel to show him where we are in project. Sal actually got nasty and dumped the blame on Operations, ignoring any of my logic, such as: How are we going to build a commercial product if we don't have working model to follow?

But after getting beat up by Sal, I earned a small victory for project management (and schedules). The team finally seems to realize that getting units to meet both milestones is impossible if we plan on using newer-revision boards. Everyone began to think out loud about ways to have hardware available to send to Sal's customer once Alpha (really pre-, pre-Alpha) firmware is ready.

Lead engineer, Will Invent announced that there were older-revision boards laying around that could be resurrected. The only question was whether there would be any problems loading the new firmware. I took the action to contact Cody Smith, our lead software engineer, to see if we have a possible solution for Sal's customer. Cody's answer: Loading firmware on previous revision boards will work. So now the bottleneck for this milestone has shifted over to Cody and his team.

I feel like I'm playing Whac-A-Mole, except my moles are the critical items that need to be knocked down or eliminated before the next critical item rears its ugly head. If the team can't react to these critical tasks quickly, the project will fail or at best be marginally successful. Cody seemed pretty confident about delivering code. We'll have to wait and see whether his confidence is warranted. I've heard this same bravado before from others, but I'm keeping my fingers crossed that Cody's as good as his word. So far, though, no one on the team has shown a particular gift for prognostication when it comes to reliable schedule dates.

July 26

Dear Diary,

My second month is nearing its end, and we still haven't seen a completed BOM for most of the boards. That means having product on time for Design will be impossible. Fortunately, engineering boards for Sal's customer just need Alpha firmware. So there's a glimmer of hope that at least one of the two milestones will be met. Our team needs some success to build on, since more engineering changes requiring new parts have been coming in and wreaking havoc on the schedule.

To help mitigate the impact that the new parts are having, I asked (nicely) the Design engineers to send me any potential part changes before the official ECO so we could order them. At this time, many of the changes affect resistors and capacitors, components with short lead times that can be purchased with few dollars at risk. No way do I want to tell Richard or Sal that we can't ship because of a part that cost less than a penny!

I updated the schedule to showcase our milestones and, as always, shared it with the team at our most recent meeting. More and more action items are getting completed, but constant design changes have made predicting end dates difficult. Dates for functioning boards are critical to the schedule, because they determine the team's ability to push the product into production. Trying to plan anything with this project team is like predicting the weather on Mount Washington – no easy task. I've started posting the schedule to the central database, along with all Operations-related items, so everyone on the team can see what's going on from a production standpoint.

ACME sent quotes to us and – whaddaya know! – their prices are very good. Peter and I had a conference call with them to talk about documentation requirements and lead times to get boards built. The best they said they can do: 2-3 weeks from a complete documentation package with all parts in hand. So we're stuck for the moment, since parts are still being ordered and board documents are in a "constant state of flux." We decided to wait another week for last-minute changes, then send all released documents and any parts received at our sister company's warehouse (we don't even have a warehouse let alone a factory).

I added the results of our meeting with ACME on documentation and lead times to the schedule so everyone can plan accordingly. Of course, waiting another 2-3 weeks for boards won't be acceptable. To minimize the wait, Peter and I shared ongoing ECOs with ACME as soon as they became available. Information on new parts and due dates also were provided. So now, once we receive the go-ahead from Engineering to build boards, everything will be in place – which ought to save us weeks (at least that's the thinking).

Learning #2: Process

Define Means to the End

Q: How do you eat an elephant?

A: One bite at a time.

Bad joke, but good approach to a project. Each "course" is equivalent to the milestones in a project. Breaking up the journey into manageable "bites" allows the project team to have small wins throughout, before the big payoff– dessert! – at the end. One way to do this is by implementing a commercialization or product-development process. Okay, so our imaginary team doesn't have one for its project, and probably won't for the duration. Do as I say, not as we did. Learn from our mistake.

Many companies use a process referred to as a stage-gate or phase-gate. "Gates" are used to evaluate progress between stages or phases, and require a Go / No Go decision – "Do we continue on, or is it time to cut our losses and end the project? – before moving through the "gate" to next stage / phase. Di Benedetto and Crawford provide an excellent new product phase-gate process, which establishes the following phases:

> Phase 1: Opportunity Identification and Selection
> Phase 2: Concept Generation
> Phase 3: Concept/Project Evaluation
> Phase 4: Development
> Phase 5: Launch

This process calls for a great deal of analysis early on to sift through the muddy waters of potential projects and select those that can provide the biggest return on investment.

Even diligently adhering to a process doesn't guarantee success, due to the many internal and external variables that teams must deal with. So the question is: How can we even begin to have a shot at success if we don't follow a new product process to ask the hard Go/No Go questions? Without this discipline, the chances of getting any product to market in a timely fashion will be darn near impossible.

Defining the means also helps calculate the labor, skills and capital needed to reach the end goal. With at least a best estimate at the above inputs, the project manager can start to understand the critical paths of the project in order to determine ways to get back on schedule. Without these necessary project ingredients, quantifying an accurate launch date with customer availability will be impossible. The end result - nothing but frustrated customers, who often have other places where they can spend their money.

Chapter 3

"That's the problem with being a Project Manager: The actual role should be Supreme Omnipotent Dictator, but it's more like Nurse Maid. I guess I don't need this sword."

August 3

Dear Diary,

Sal is sticking to his ship date of the 15th for one of his more important customers. For this milestone, we've managed to get the hardware, but we're waiting for software – or at least an installer that will add it once it's ready. Everyone knows this is the current bottleneck. Rule of thumb: Software is always the last thing to arrive. This is due to the vast number of features that customers demand, software fixes for hardware problems, "feature creep," and the apparent inability of anyone to accurately forecast software tasks. To top it off, we don't have enough "people power" to get the job done on time. (My "little voice" tells me that appropriate staffing is going to haunt us throughout this project.)

No one seems concerned (at least for now) that, day after day, no installer arrives from Cody's team, which is in keeping with the personality of this place. Meanwhile, Peter has been staying on top of our suppliers for parts and cost reductions. He's ended up becoming a BOM Warden, constantly finding problems that would kill us once we're in full production. It's a very dirty, stressful job, but Peter's up to it, and we all ought to be glad he is. He's been scrutinizing all the ECOs so the transition to ACME will go as smooth as possible – not the same thing as saying things will go smoothly.

I nicknamed him Rain Man (after Dustin Hoffman's character, Raymond Babbitt), because he's always rattling off part numbers for components and boards. Peter's ability to juggle tasks is world class and he doesn't drop any balls – and our balls are made of glass! A side benefit is the amount of fun we have working together. I've noticed humor is playing a huge role in helping the team remain at least somewhat sane as we continue to do battle with all the problems.

We just wrapped up our most recent schedule review, and the projected end date is … well, it's out there. It looks like the only way to make up time is to roll the dice and build boards for both design certification *and* customers at the same time. We'll have to take two serial tasks that originally depended on each other and collapse them into one. It'll save four to six weeks – and now for those famous last words – if all goes well. Did I say this is a very risky way to improve time-to-market? But we really have no choice at this stage. If the boards work, we'll have the potential to quickly satisfy customers. If they don't, we'll be adding to our pile of scrap boards and project delays. And I'll be dusting off the ol' resume and scratching out a new cover letter or two. Wonder if I can use Richard as a reference?

I changed the schedule again to reflect this parallel approach – and watched in amazement as the end date got reeled in by five weeks. If only project management was always that easy: Change a schedule, pick up a month and change, and everything and everyone involved just follows along and gets the job done. But my excitement was short lived. After closely reviewing the schedule, a chill ran up my spine. There are so many things that could derail this train. It's a matter of which item will eventually cause the horrific crash. If past performance is an indicator of future activity (sounds like an investment firm commercial), we're in grande trouble. Not only have our estimates to complete design tasks been off by months, the product has never fully worked. Just amazing. But the schedule … hey, it looks good.

August 10

Dear Diary,

This week's team meeting focused on the new schedule and how much confidence everyone had in achieving this parallel effort. Engineering is convinced the next revision of boards will work flawlessly and adhere to the new standard. I hope they're right. One thing that irks me is that no one in Engineering has ever worked late or weekends, even as the schedule has slipped by months.

My personal belief is that you do whatever it takes to get the job done – period. It doesn't matter how much extra effort it takes. And here we are as a team, taking a huge chance to make up time that could have been lessened if others on the team put in some extra hours. If I were Richard, I'd be kicking some donkey – and I don't mean donkey – about now. That's the problem with being a project manager: Supreme Dictator is not part of the job description. Too bad.

Sal sent me an e-mail inquiring about firmware for his customer's units, so I called Cody to see what was going on. He said the software would be ready next week, and I'm not sure Sal is going to be happy with that news – he wants it now. So I called Sal and shared the new due date and "in parallel" update schedule. He didn't like it and was frustrated with me, even though someone else "owns" the task. Project managers are messengers of good and bad news, so I guess the grief Sal is giving me probably should be expected. I highlighted Cody's action item, which will remind me to call him frequently to check on things and push this task along.

Here's a bright spot: The team is making actual progress, largely because specifications haven't changed in the last two and a half weeks – a long time for this project. But then Sal started to talk to an ODM (Original Design Manufacturer) who appears to be ahead of us. On top of that, the ODM's product costs are lower, and they're willing to talk about a partnership. Time for yet another change in direction!

And we're off. Operations has been told to quickly order product from Sal's ODM, with delivery expected in eight weeks. This is adding another complexity to the mix: Four different locations for the team, a potential supplier in the Far East, a project that's way behind schedule, the RoHS/WEEE directives, and so on. Boy, this project is way out of control. Defining the final destination hasn't happened, so now we have multiple final destinations – our product, the ODM's product or some combination of both. We haven't even determined if RoHS/WEEE compliance is going to be required. Time to put the ol' diary down before I write something I'll regret.

August 17

Dear Diary,

Here comes the 15th and – whoosh! – there goes the 15th and nothing has been shipped to Sal's important customer. No matter for we can always change the schedule. We now have a target date of September 1st for all parts. In essence, Peter has been trying to force Design to meet a fixed date. If all parts are on hand, the bottleneck moves to Engineering to complete the final board layout. It's what an old boss of mine called a "Forcing Function" – something that blows through everyone's personal agenda and forces them to act for the sake of the overall project.

Even though I know achieving this date is going to be difficult if not impossible (since one of the board designs isn't finished), I'm going along with Peter's date. The schedule and its parts due date clearly shows the impact a late design will have on our ability to build boards and a finished product. Putting a date in, that's probably not attainable, isn't something I particularly want to do. But a certain level of desperation to push the team is causing me to try anything to set a fire under members.

The frequent sharing and communication around the schedule has, in fact, enabled the team to make progress. But the pressure from Richard has forced the team to disintegrate further into functional individuals. Everyone is focused on their small part of project, hoping to pass the bottleneck off to someone else. It isn't pretty, but it's moving the needle. It's human nature to survive – even within project teams – and no one wants to be "the critical path" when Richard starts asking questions during meetings.

Call it Project Hot Potato and it works like this: Even if the result of your particular effort is less than stellar, pass whatever appears to be working on to Operations so we can procure remaining parts and plan for production at ACME. And about that firmware release for Sal's customers: It's delayed one more week, even with my constant cajoling of Cody.

August 26

Dear Diary,

With the end of the month approaching, lo and behold there was a glimmer (albeit a small glimmer) of hope that all of the boards would be placed into production soon. It turns out the light at the end of the tunnel was an oncoming train. It was just more wishful thinking. We don't have a chance of achieving this lofty goal. We're still at least a month behind schedule on one of the boards, and no one will admit this to Richard. Everyone's so busy they're starting to ignore the schedule and pushing ahead with their tasks. Meanwhile, we've been getting constant pressure from the gang at ACME, which is anxiously awaiting consigned parts for builds. So they're not in a real good mood.

I recently started sending ACME documentation on boards released into our MRP system that have little probability of last-minute changes. My counterpart at ACME is a real firecracker and keeps me honest. Within hours of receiving our documents, her team noticed some inaccuracies in the BOMs and pushed us for answers. (Too bad I couldn't have her tackle some of our design issues.) Given the two-to-three week lead time from complete documentation package and parts to assembled boards, I wanted to hit the ground running and get through these deliverables well ahead of the engineers. Like I said before, waiting until all the parts arrive before sending documentation will just cause more delays that we can ill afford.

It's time to remind myself that project managers have little control over most project tasks. When there are ones I can control, I try to attack them with all my might. Why? It's a good idea to lead by example and set high expectations for others concerning what's required of every member. So I spent most of today getting all the documentation out and feverishly sending e-mails to Engineering to address the BOM and part specification issues raised by ACME. No way am I going to be the bottleneck and let my tasks get in the way of progress.

Oh, and about that firmware release for Sal's customers: It's delayed another week.

Learning #3: People

Humor Helps

Personally, I've never read any formal business text that mentions humor as a means to achieve business success. But in my experience, it helps. And there are many not-so-formal resources available that provide guidelines for including humor in everyday business interactions. Humor is a great way to foster teamwork, especially during those tense moments when Murphy's Law rears its ugly head. A well-timed joke lowers the tension, improves communication and brings a team together to tackle the next problem – and there *will be* a next problem.

Even though no one on our team has read *Humor at Work: The Guaranteed, Bottom Line, Low Cost, High Efficient Guide to Success Through Humor*, by Alpern and Blumenfeld – a recommendation from my younger brother Joe, who's a Certified Laughter Leader. Every team member, even Richard, has used humor to ease tension. Either by accident or by design, virtually every manager of every project team I've ever been a part of has also used humor to improve team work. There's an intuitive understanding concerning the value of humor. You might want to pick up the book for a more analytical approach.

Humor works to prevent team members from choking each other as, day after day, the inevitable problems take their toll on schedules, budgets and relationships. Sometimes a light moment can be used to "nudge" someone who's gumming up the works. But be careful: Humor isn't always received in the spirit in which it's delivered. Operations has created nicknames for most everyone, even each other. Richard is the BGH (Big Giant Head), Sal is Oxy (as in oxymoron given his marketing intelligence), and Dean is The

Dream (based on a compliment paid to him from a female co-worker). My nickname is Superman. I gave it to myself. That doesn't stop everyone else from calling me – well, this is a "family" book, so let's move on.

Yeah, nicknames and certain other popular forms of humor are immature; something we should've outgrown back in grade school. But it sure is enjoyable. And, hey, experts say it's a good thing. Dilbert cartoons alleviate stress and frustration – and our project team and the circumstances they face (and create) resemble many of the daily creations by Scott Adams. It's as if Adams has a crystal ball – along with all his years in Corporate America – that looks in on today's companies, and he sketches the next day's cartoon as he observes us all in action.

For me and many other project team members, one of the biggest benefits of working on a team, is meeting new people and gaining new friends. Humor certainly helps bring these friendships to fruition.

Are We There Yet, Diary of a Project Manager

Chapter 4

"They say good Project Managers step on toes. Fire up the music – it's time to start dancing on some Dr. Martens!"

September 4

Dear Diary,

September 1st came and went, and although we haven't received all the parts, many are resting comfortably in the warehouse. There's a chance production can start for some boards while we wait for the last board designs to be completed. The missing BOM ended up being the tent pole (really the skyscraper) in the design process – the last item needed to build pre-production boards. The project schedule clearly highlighted this fact, something that many on the team – staying firmly in character – have barely noticed.

One other concern: We don't know what surprises await for this board around the supply chain corner. For example, are there any long-lead parts that will further delay the schedule? Peter gently asked the design engineer (for what I think was the 10th time) to send whatever he had on boards so he (Peter) could start quoting and ordering any new parts. Operations is being as proactive as possible to improve time-to-market.

The engineer responsible for this board, a fellow named Isai Goodenough, should have been shot a long time ago, but someone went and made that illegal. This fellow manages to survive under the corporate radar, but always seems to show up at just the right time to say just the right thing. Every project I've ever worked on has someone in Marketing, Design, Manufacturing, or even Management that everyone knows is going to be the major reason for a delayed project schedule. In fact, if there was a "Most Likely to be The Critical Path" contest, they'd win even before the swimsuit portion.

This species seems not only to survive, but to breezily move on to their next assignment, which often is the company's next debacle. This phenomenon continues to amaze me. What's more unbelievable, is that they don't (or won't) recognize that they're a problem. Over the years, I've tried to compensate for this species so it doesn't bite me. In this instance, it bites me hard. Good thing my tetanus shot is up to date.

From my project manager perch, my team's ability – that is, its fundamental inability – to meet commitments has become readily apparent. If there was time, I'd analyze some of the current schedule delays and develop a standard "delay deviation" (also called a slippage factor) for everyone on the team, including Operations. Then my schedule could include a probability of success based on the empirical rule.

Adding one standard deviation of additional task time would yield a probability of success of 68%. Two would yield a 95% probability and three 99.7%. Since there's little spare time to tackle this issue, I went ahead and estimated "inherent delays" of the team members and prepared to add these figures to future schedules. I should be able to show what is realistic given resources, skills and tasks, and then figure out ways to improve the schedule. I've learned that loading up a schedule with optimistic numbers that match someone's idea of an acceptable end date does no good. Gotta "man up" and avoid that trap, even if it means a decline in my popularity polls.

September 10

Dear Diary,

Second verse, same as the first: The team slipped past its August dates and will soon fly by its September dates. So far we're zero for two, missing milestones for delivery of the customer trade trial and the new boards to Design. The Operations team believes boards will be ready by late October or early November, based on parts availability and late board design. We keep pushing and smiling.

During our team meeting, VP of Operations Manny Factura apologized for all the recent craziness. We're used to it by now, which might or might not be a good thing. I mean, what would a project be without some insanity? But maybe enough is enough. This project resembles the movie *One Flew Over the Cuckoo's Nest*, with Richard starring as both McMurphy *and* Nurse Ratched.

Weeks ago I offered to be the project manager for the entire project, including design activities. The request went nowhere with Manny or Dean. Good project managers step on toes, and it's time for me to start dancing on some Dr. Martens. In order to better predict the probability of meeting key dates, I need to get a list of milestones from Design. After all, you can't build something that hasn't been designed. Once I receive a green light and the milestones from Dean, I'll consolidate the Operations and Design schedules and determine when product will be available. I'll put this item on our weekly action item list with Dean as the owner. We'll see if this constant reminder gets him to react. (First we'll see if he agrees to provide the information at all.)

No milestone document from Engineering yet. About all we can do is keep pushing the engineers to provide documentation so we can build some boards. All board documentation must be released into the MRP system, so eventually we can leverage it for production planning. Meanwhile, Peter created a complete product planning BOM (what a guy!) so he can manage on-hand inventory, ongoing orders, receipt of parts and eventually receipt of completed boards.

What our team lacks in overall discipline, Operations does its best to make up for. But a team is only as good as its weakest link – and in our case, we can point the finger at Richard, aka, The Big Link. After all, everyone on the project works for him and takes his direction (or lack thereof). In the end, it doesn't matter if we track progress on a detailed schedule or have a separate complimentary spreadsheet with action items that we review daily. The team could be doing all this and still head in the wrong direction, or at least not go about the trip as efficiently as possible. Overall direction-setting is; plain and simple, a leadership issue. And Richard's leadership capability or Cpk is negative.

September 20

Dear Diary,

News Flash: The parts shortage list is shrinking, and Operations' prediction that building boards will start in October is looking better and better. It actually matches the schedule – a rare but satisfying experience. There are just a handful of parts we have to monitor. The last board design is still in process, and Goodenough's manager protects him like a mother hen. In fact, all communication between Isai and Operations goes through his boss. So for now, Richard has been replaced as the critical bottleneck. My schedule, the action items list, and pretty much every conversation I have, centers on when this board is going to be ready for prime time.

During a recent project call, Richard pushed Design for test plans – another critical part of any commercialization effort that we've given little attention to at this (or any) stage of our project. So he's asking the skunk works Design team to document something that it has no primary knowledge of and no business doing. The icing on the cake: Richard questioned the team further about testing, and someone admitted they're missing critical test equipment (nothing like waiting until the last minute).

I didn't understand much of the conversation around testing, but I know almost all of the Design guys were diligently taking notes as Richard talked. He reminded us he's been in their shoes before, and offered some great insight into how to test a product and its components. What was astonishing to me and a few others was that the engineers, once again, got away without documenting their test plans. On the upside, the missing critical test equipment was ordered after the meeting and is due by the end of next week.

Operations also got pushed for due dates on boards. Our best guess, based on the updated schedule and on-hand inventory, was early October to late November – how's that for precision? And that estimate didn't include the last board that's still being created. Dates are based on two-to-three weeks from ACME receiving the last part. I hope the upfront work we did with ACME will improve the date or keep us on track if one of the delivery dates slips.

September 26

Dear Diary,

Peter and I are traveling to ACME next week, so most of our time lately has been spent preparing for the trip. Arrangements have been made to send parts over there. The goals are to set up a separate inventory location and answer questions related to our previously sent documentation. This is our way of making sure that when the last part arrives, production can be set up and boards can flow through the factory as soon as possible. We don't want to have any more delays in getting these boards. After complaining about others' inability to meet the schedule, we can ill afford delays on our part.

There have been some amazing requests made of Operations during this project, but one late in the week took the Olympic Gold Medal. Dean asked if we could build the board that was still being designed by Isai Goodenough, because Engineering was concerned about problems they might encounter in pre-production. When I said we'd be more than willing to build the board once its design was complete, the BOM was released and parts were ordered, Dean was shocked. You can't make this stuff up.

We've communicated the project's status through schedules, action items, e-mails, phone calls and project meetings – and yet the HEAD OF DESIGN is oblivious about the delays in DESIGN. The only thing that could cause this magnitude of disconnect is separation of the hardware and software design groups with no communication between them. As the weekend neared, I got into an unhealthy game of Brain Ping-Pong (there is no healthy game of Brain Ping-Pong) to figure out how to address this communication vacuum. Maybe Dean is sleep working and really doesn't understand anything we say in meetings, even though he appears to be engaged.

Learning #4: Process

Specify the Means

A best-case scenario has a project team breaking up its project into more manageable mini-projects, then detailing each task – with duration, start/finish dates, owner, and predecessor and – most important – what could go wrong. This is the detailed project planning that must be done with realistic data – people, equipment, software, etc. – in order to be successful. If data of any kind is missing, the team must address this as soon as possible.

This is usually the point where some manager will mention the use of Microsoft Project or another formal means of tracking progress. But any project is only as good as the data – garbage in = garbage out. Task management is more than knowing the task owner and the percent of the project that's complete. So what can be done to drive a project towards more accurate and realistic data and attendant tasks?

Focusing on pre-development activities enhances the probability of successfully commercializing a product. *The Project Management Manual* [Bowen] breakdowns each project into three sections:

- Define & Organize the Project
- Plan the Project
- Track & Manage the Project

As you know by now, the team profiled in our diary doesn't have a formal development process or an overall project manager (despite my offer). What it misses most as a result is the "Work Breakdown Structure," which is part of Bowen's "Plan the Project" section. Major delays are being caused by forgotten or omitted tasks and a lack of the resources needed to complete them. Where volume is the key determinant of resources required to plan factory capacity, a breakdown of the work is what's needed to calculate resources, such as FTEs (full-time equivalents) to meet the target date.

This basic target date exercise can be completed using Microsoft Project or just a simple calculation – i.e., the total hours needed to complete tasks divided by current resource hours/weeks/months. If the answer is past your targeted end date, either more resources must be added or some tasks must be outsourced or eliminated. Something's got to give. You can squeeze only so much work into a fixed time frame at a particular resource level.

Resources don't necessarily have to be internal. Maybe a supplier or contract house can help provide additional capacity. An example regarding our project was the considerable help that we received from ACME Inc.

Many project teams, in the beginning, don't take the time to analyze commercialization resources at this level of detail. Doing it early on is difficult, since there are many variables that won't become evident until later in the project – such as understating tasks, lacking necessary skills to complete tasks, new part lead times, competitive actions, and the dreaded specification changes or "feature creep." The more successful projects make these calculations as a matter of course – as diligent with this as with constantly "running the business case" to make sure the project still has a positive NPV.

Are We There Yet, Diary of a Project Manager

Chapter 5

"As Project Manager, I must understand the unique skills each team member brings to the project and what it takes to have them share those skills for the success of the team. Or I can go play Tiger Woods PGA Tour 07 with my sons."

October 3

Dear Diary,

Peter and I spent the week at ACME helping set up the facility to produce our circuit boards that are (a mere) two months behind schedule. We prioritized boards based on our critical path, customer requirements and parts availability. I felt right at home with my peers at ACME. There's no doubt we can rely on them for help addressing many of our project needs.

Over time, I've learned that successful project management depends on bringing partners into the team that provide complementary skills and additional bandwidth. ACME is just such a partner.

The kick-off meeting included a discussion about whether we're designing to the RoHS/WEEE directives. How do you answer when Marketing has provided a total of zero direction? It's clear to me, that this will end up being a requirement, since a majority of our products are probably going to be sold in Europe. My project management role will definitely expand to include making sure products meet these directives. So as part of our tour of ACME, their production manager showed me how they intend to handle lead-free soldering to satisfy RoHS.

As we all went over the BOMs and board layouts, we discovered parts that were wrong – for example, parts with incorrect pad geometry. On top of this, critical parts we need for a board got mixed up with another shipment and ended up in – Italy. Murphy lives! And he's killin' us. Engineering gave us the approval to build boards without these parts, and with other parts that had been incorrectly specified. The engineers can place the missing parts with the samples they received. The parts in Italy (enjoying some red wine and pasta?) are going to be sent back. We think.

Meanwhile, Sal hasn't attended our last few calls because we've missed our promised customer dates. He isn't happy. He doesn't talk directly to the engineers, but communicates with them through Operations (lucky us). Then he's surprised by how far behind we are in producing units. Someone famous or at least someone smart once said: "You can't produce something that hasn't been designed." (OK, it sounds like something someone smart would say.)

Oh, and about that missing board BOM: We still don't have it. Operations has ended up in the role of Ernestine the telephone operator (made famous by Lily Tomlin) so our various functions have a reliable communications channel. At this junction in the project, I've created an even simpler schedule to better focus team on critical short-term tasks.

Lately I've been thinking that I have to do a better job understanding the unique skills each member brings to the table and what it takes to have them share this for the success of the team. OK, it's a little late in the game, but I haven't been able to catch my breath most days. Each team member has different communication filters too, so learning more about this diverse audience is important if I'm going to help drive changes to enhance communications.

Are we having fun yet? I have to respond with an emphatic: "Yes." I've seen problems in most every project. The trick is to fix them quickly and move on – it's the difference between success and failure. The longer it takes to get a product to market, the bigger the cash hole, the more susceptible you are to competition, and the harder it is to fill the cash cavity. So we've got to address problems expeditiously so we can launch this product. I have to say, that despite all the problems and my monthly bill for antacids, I'm enjoying this. Call it what you like – just don't call the psyche ward.

October 10

Dear Diary,

ACME really pulled out all the stops. The expedited boards have arrived at our engineering facilities. But no one is checking them. In the meantime, Design has been working feverishly to incorporate another company's solution into our system to meet a Richard deadline. If this craziness isn't enough, during the team meeting Richard dictated a new design approach. He wants modules that can be shared across both low and high-band products. Makes sense, but we're well into the current design and it's a little late in the day to throw something like that into the mix. The project has hundreds of thousands of dollars invested in parts, bare boards and assemblies that are being built at a premium. Now this? The same person concerned about product cost, now is making a decision that could cost us hundreds of thousands of dollars – and possible delay the project even further!

Right after the meeting I started to put together a spreadsheet with details around the proposed design, because the new approach is just as confusing as the original. It's taken me months to learn the nomenclature of the current products and boards, since engineers use different terms to describe them. I need to get up to speed on the new design as soon as possible.

I also want to track our progress and force the engineers to provide realistic due dates so we can understand the expected launch dates. Knowing how well the engineers have met (really missed) past due dates, gives me a chance to modify their estimates to improve schedule accuracy. Hopefully it will provide the project discipline that was lacking before. I've managed to gain some respect from team members, so I can expand my reach even if it means not being liked – mostly by Design and Marketing (hey, no great loss). Being a project manager is not a prelude to winning a popularity contest.

Richard gave the team a week to assess his new approach and provide feedback on its feasibility and a timeframe. It's hard for me to formulate a clear schedule, since the bulk of the work and analysis has to come from the engineers. At least he's being reasonable about a new path forward. We'll see if this lasts.

Sal awakes! He sent out a blistering e-mail toward the end of the week stating: "We need to graduate from an engineering-run company to a sales/customer-run company." He's absolutely correct, too – as long as you reverse sales and customer. The problem is that engineers and technology run our company. Then when we don't have a solution, Sales creates them out of the blue.

Everyone usually points to Operations when things look bad, but not everyone truly knows what it takes to get a product to market. A day after Sal sent his e-mail, he started talking about *two new* products. If you're keeping score, here's the situation:

- Project team is six months behind schedule
- Richard just introduced a new modular approach

- Sal just created two new products
- No decision on RoHS/WEEE certification
- Isai hasn't completed his board design and BOM

Back through the looking glass again!

October 15

Dear Diary,

We all spent the first part of the week trying to figure out what will be "common" between the current designs and Richard's new modular designs. Operations wanted to be prepared for the weekly team meeting, so we've been in constant communication with the engineers. The meeting began with everyone ready to share their concerns about the new design. To no one's surprise, Richard ignored these concerns and said he wanted new boards designed, unique parts/bare boards procured, and board assemblies built so product can be ready by the end of November – he means this November.

So boards originally scheduled to be designed and available by December are somehow going to be part of a November miracle! I always knew our engineers were imaginative, but now Richard is part of the mix – this is going be interesting. How much do you want to bet that he'll give us a hard time when we ask permission to buy unique parts and bare boards?

The benefit of the new concept (yes, there is a benefit) is that the product portfolio will be documented and used as our "Design Bible." In fact, over the weekend I created a product family plan and shared it with the team on Monday – yet another communication tool to keep us on track. But this one does help me to sort through the quagmire of board names and product models.

Also on the bright side, this mid-course correction is a new opportunity to make things better – at least that's the positive way to look at the situation (remember, I'm an optimist). Now it's time to rework the schedule and see what the new project dates look like, making sure to highlight all of the major milestones early and communicate them frequently.

I pulled together a one-page document with the following "Project Scorecard":

- Design Readiness
  - Board design
  - Software design
  - System integration
- Marketing Readiness
  - Customer trials
  - Shows
  - 4Ps – Product, Price, Place, and Promotion
- Quality
  - Issue resolution
  - Regulatory testing/approval
- Production Readiness
  - Parts
  - Tooling
  - Test
  - Assembly
- Launch Date

Nice, eh? The project schedule is supposed to drive dates for the "Project Scorecard." Ideally, this document will be shared throughout the organization and make project delays obvious.

Everyone will agree to a first revision, including a code that would be as follows:

- Green – complete
- Yellow – in process
- Red – schedule delay

Any red causes a call to action to quickly address the situation so things stay on track.

October 25

Dear Diary,

Richard met with Operations to find out how quickly we can build the new boards. Richard likes to meet with us because we come prepared with eye-catching, color-coded spreadsheets, project schedules and presentations. Since we've had this discipline from the beginning, Richard always expects this level of detail from us. No other group in the project team provides anything along these lines. Richard never expects it from them either.

In a way it's good that we can provide actionable data in order for Richard to make decisions. In another way it's bad. Potential problems are usually presented by us even if we're not responsible for their resolution. So Richard ends up looking to us to fix problems in other organizations where we have absolutely no authority. Bottom line: As much as Richard likes to lead us around on this National Lampoon Vacation, he despises confrontation of any kind, so no one gets the ass kicking they deserve.

We showed Richard our schedule, which lists board deliveries one to two weeks later than he has set for availability. Not an especially happy moment in the meeting. After having been burned by overly optimistic estimates, and finally getting a handle on team members' ability to meet due dates, I've made sure my schedule showed the stark reality of the new strategy. There was a candid discussion to try and come up with ways to save time. The only answer is for Design to shorten its time frame for board layouts.

To his credit, Richard agreed and said his next step would be to pressure Design into getting things done sooner. But most of us see this as a feeble gesture; since Design is stretched so thin it'll be lucky to get within a couple months of its target date. Back at the ranch, there was great news for Manufacturing: We got a green light to buy the initial set of parts and boards for the modular design.

As painful as it's been to do a design change at the 24th hour, we're also excited – reenergized – by this new approach and challenge. Modular design halves the number of boards we'll need to build. In addition, tooling and set-up costs can be allocated across a greater number of boards. This will decrease unit manufacturing cost, which is critical.

Fundamentally, we now have the ability to configure a variety of products based on customer requirements with this series of boards. We can also take advantage of this to begin specifying RoHS-compliant parts, since the engineers are basically designing a whole new set of boards. So there are many silver linings in modular approach. But we still we don't know the number and size of the clouds. There have been so many rainy days around here, I feel like I'm in England.

Learning #5: Process

The Devil is in the Details

A catchy phrase and also true. But many project teams never take time to uncover the "devilish" details and end up with a project that's behind schedule, over budget and ready for the scrap heap. The details we're talking about here shouldn't be confused with accurately documenting project tasks. We're talking about understanding the "what ifs" surrounding potential risks – those tasks that will be difficult to anticipate or achieve, and might require an alternate plan.

OK, you say, but how do you plan for the unknown or the unexpected? The key to spotting the "devil" is knowing that every project involves risk. What many projects lack is a thorough assessment of potential issues from competitors, customers, economic conditions or regulations that can derail the program. Create a Risk Management Plan at the beginning of the project that includes preventive actions to mitigate risks and propose contingencies in the event of a failure.

Certainly within their fields of expertise, team members can handle problems that arise – a spurious interrupt on a board, an inaccurate part specification, a mix up in shipping, a problem in the production of boards. What's often missing is a commitment to "expect the unexpected" throughout the program, and a reasonable – if not well-defined – plan of action if and when these unexpected situations occur. Production can hire temporary workers, build utilizing full capacity, and transition from one to two or three shifts to meet fluctuations in demand. But what happens when Design can't handle the sheer volume of work it has? How do you plan for and correct this situation in the middle of the program?

Successful projects have a better understanding of risks, a more thorough plan, and early preparation for "devilish" events. Since upfront planning of any kind is non-existent on this project, tackling any kind of project crisis invariably delays things further. Our team just doesn't have the resources and spare capacities – not to mention the will or the focus – to understand situations that could develop and contemplate contingencies to correct them.

Successful teams, especially the ones with experienced members who have worked together previously, nourish the ability to handle the "what ifs." I've been fortunate enough to be on these types of teams. When problems arose, there was no panic – just a workman-like attitude to get at the root cause of the problem, fix it and move onto the next challenge. In our case here, most of the team members haven't worked together, so cultivating this skill is taking time. The best way to handle "what ifs" would have been to anticipate them in the beginning and plan for them before they happened – but, of course, that was something they didn't do.

The only area where they planned for "what ifs" was hiring contract software engineers from India to support Cody. At least Dean understands the capacity of Cody's team to drive all of the programming requirements, and realizes early in the project that he needs extra help.

Chapter 6

"Project Managers cannot be chameleons. We must always present the facts, and nothing but the facts, even if they are unpopular. This, of course, is what long-term disability insurance is for."

November 3

Dear Diary,

Getting Richard to approve unique part procurement was a bit of good news, but the hits keep coming. Yes, we finally – finally! – received the remaining board BOM and documentation on the last day of October from Isai. And the firmware installer for Sal's customer has been loaded on units – only two months after his initial request. But ...

If I'm going to be honest, I have to give myself a failing grade for keeping pressure on Cody. There's no excuse for taking my eye off this task simply because of the chaos that's inherent to our project. After all, it's my job to keep my eyes on all tasks, and not let anything slip through the cracks – especially as it relates to tasks that directly affect customers, the people I advocate for on a regular basis. (Not a bad mea culpa, eh? Okay, now I feel better about myself.)

Meanwhile, my initial enthusiasm over our new modular design approach is waning. Things that caused delays with the previous design effort are causing delays with the new design effort. The development engineers keep changing the design on the fly, so we can't get firm BOMs to start purchasing unique parts.

This is the critical path: We need to understand what the new design entails or production will continue to be at a standstill.

Once again, the "Powers" expect product in an unrealistic time frame. Richard is leaning on the team about the cost of our new build, even though he previously gave us approval to purchase parts, boards and contract assemblies. For the first time, Operations is getting visibly frustrated – assurances from Richard, Sal and Dean aren't being kept. It's the same old-same old story of pushing problems downstream so the Manufacturing group ends up in a no-win situation. No project schedule is going to prevent the tidal wave.

Instead of waiting for Design to provide parts with the appropriate documentation, Operations has taken on the task. In fact, maybe for the first time ever, Design actually was told to work over the weekend to get BOMs complete – there is a God! It's almost impossible to track parts, cost, inventory, production builds and engineering changes without releasing a bill of material into our MRP system. That's why Operations keeps pushing for documentation. Add to this, the fact that Richard keeps asking questions around inventory, board and product costs. The only way we can efficiently get this data is through the production system.

Just to keep everybody from heading for the exits, we received documentation for one of the boards (five more to go) and Peter is getting pricing and delivery dates. Our greatest concern in hitting our end date is lead times for new parts. Some of us insist on remembering that our previous procurement effort took almost four months. The reason no one noticed at the time, was that delays in Design were greater than part lead-times, so everything arrived simultaneously for the production build. So from Richard's standpoint, there was never an issue with the supply chain even though he and Design caused many delays. Not the type of memory that can be easily erased by the project manager when his schedule states the opposite.

Sal started his customer trial campaign again. My e-mail queue is filling up with his none-too-subtle messages about the critical importance of hitting our dates. Hey, thanks for the heads up, Sal ... I totally agree. But once again we're asking for the impossible with an artificial target date set by Richard. I've been keeping my fingers crossed that moons align and we somehow manage to pull this off. At this point, though, I've decided to develop a new schedule to track the progress of new boards only, since there's no way the team can focus on the big picture.

November 10

Dear Diary,

Tick-tock. Instead of waiting for the perfect bill of materials, Operations decided to buy everything unique to the new design. There's a risk of course, in procuring items that might not be required in the final design. But we couldn't risk missing build dates because a single part is unavailable. Operations is taking matters more into its own hands to help make sure this project will be successful. These gambles aren't being shared with Richard (shhhhh!!!). We aren't stupid, and besides we can't afford to waste time on another conference call to review costs. Plus, we don't want Richard to add yet another roadblock to our journey.

For the record, this is one of the rare times on this project when I have confidence in our ability to achieve an end goal. Why? We have an experienced team tackling the right activities required to get boards built. Even though the engineers have been the focus of our frustrations, they're really working hard and the ultimate success of the project squarely rests on their shoulders. One good thing about Richard dictating the date is that team members can't hide behind functional walls, coming out only when it's convenient. The whole team is being stretched so thin, at times we're close to breaking.

Miraculously, we're on schedule – on our new schedule that is – with our major deliverables: schematics, layouts, BOMs, etc. No idea at the moment about software status. There are plenty of "cooks" in the programmers' kitchen. It'll be interesting to see how well it works. We're also waiting to see what land mines await us once unique parts are ordered. I did a quick assessment and found out some parts have lead times of greater than eight weeks – that's not good. It looks like Operations will have to find a way to get these parts within the next two weeks in order to stay on schedule.

Meanwhile, it's become clear to me that, in addition to the project schedule, a separate and more detailed schedule is needed for the supply chain. If we can force the engineers to release parts into MRP when they're included in the design, then reports can be run to determine critical path items. Faced with an "engineering centric" organization that oversimplifies the supply chain, maybe I should've emphasized the importance of the supply chain and lead times earlier on. Maybe someone would've listened. Maybe pigs will fly! Oh, well. Waiting for Richard to approve parts procurement has added to the chaos of our supply process.

November 18

Dear Diary,

Once more, with feeling: Operations has set firm due dates for parts to keep us on schedule and no one's laughing (to our faces, anyway). We're just waiting now to see if the procurement effort yields the same result. I updated the schedule with the best possible dates to meet Richard's goal. Then I shared this information with Operations, and here's where I got a bunch of laughs, not to mention comments about the chemicals. I must be ingesting to think there's a remote chance of having product available in this time frame. Funny, guys.

I've even been getting grief from ACME, which has formulated a similar assessment of my mental state. Most of the concerns center on getting documentation from Engineering (which, like every other team, is overloaded) and the unknowns around this new modular design. Someone suggested that I'd fallen into the trap of making the schedule match Richard's end date instead of telling the truth. I don't know. The Operations-related activities are a stretch, but they're achievable. The real concern is with overly optimistic dates for Engineering and to some extent parts. Maybe I'm hoping that Richard's edict to engineers will force them to put in the necessary effort to meet his commitments.

Two days after the last team meeting, Design came up with three new boards required for product verification. Fortunately, they're composed of a series of connectors, but it'll be interesting to see if we have any lead-time issues. To absolutely no one's surprise, the vaunted modular design that was going to limit the number of original boards is actually no better than previous approach – maybe worse. Even though there are fewer major boards, there are more assemblies that consist of a base design with minor parts changes.

It turns out that we have maybe 50% more assemblies than before – a sad fact that wasn't visible when the modular design concept was shared with the team. Yet another minor overlooked detail that could cause huge problems in production. More assemblies to manage means more BOMs to maintain, more parts to inventory, more products to schedule, more tools to buy, etc., etc. It's not a big deal for Design – the way they see it, what are a few extra versions of a board? Just a handful of resistor, capacitor or connector changes, right? But for Manufacturing it's another variation that could cause us to miss a customer ship date. Not that we'd be breaking any new ground if that were to happen.

November 28

Dear Diary,

Before the month ends, I contacted every engineer and painfully went over each and every task and status to improve the latest schedule. This same step was also done for ACME, Purchasing, and our other suppliers to make sure the schedule is based on the best possible data. I'm now blind, hoarse and headed for the little boy's room. Where I can expedite a task – for instance, with ACME if overtime and weekend builds are utilized (and documentation can be sent ahead of time) – the shorter duration is included. Nothing like trimming the fat.

Sure, some of these time savings will actually be used up by other project issues, but overall they'll minimize or negate delays. So in the span of a week, I've exhibited schizophrenic behavior – first attempting to match Richard's due date and then awakening out of my trance to get at the truth. Given my responsibilities, I really can't be a chameleon. It's "just the facts, Ma'am," and nothing but the facts, even if they're unpopular.

So, it's now the end of November, and even though there's been a Herculean effort by our team, our goal of having pre-production units for trials at month's end isn't going to be met. We released and ordered well over one-hundred new parts, prepared to build ten new board assemblies, and handled an overabundance of design changes. Even so, there isn't enough time to account for some critical part lead times. Now the best case for the board build is the end of December – Happy Holidays to me! – if we have no more surprises.

Given the circumstances, I think our team has done a pretty incredible job. The pressure of meeting the deadline has actually fostered more teamwork. All of us are starting to understand how to depend on each other's strengths to overcome obstacles and get the job done. Even though it's taken almost six months to get to this point, there's a greater sense of team camaraderie. And that can only help moving forward. Of course, sales are nice, too. I hope there's not too much more "forward" to have to go to.

Best News of the Day (Week)!: Those cobbled units that we sent off for certification somehow passed. And we learned our company is actually ahead of its competitors – a fact that, frankly, shocked Operations. Maybe it shouldn't. A new project is a new project, and the other guys almost certainly are facing the same dynamics and dilemmas that we do. But the team really needed to hear some good news, and that did the trick. We perked up and started to plug away again.

My attempt to create an overarching "Project Scorecard" has been put aside, but that's not that big of a deal now. A simplified schedule for boards and team action items will be our communication vehicles for now. Much to my delight, a majority of the team members now see the need for documenting progress.

The reason for this change of heart: People started noticing all the critical tasks that were being overlooked. Imagine – professionals agreeing to document what needs to be done and team members joining a formal process so they can gauge what's going on and what's missing. Why would anyone think that bypassing these types of activities wouldn't have a devastating impact on project? It's a mystery …

Learning #6: People

Old Habits Die Hard (if at all)

It's one of those nagging facts of life: Unless there's some catastrophic event, people don't hastily change habits. Eventually we learn and adapt, but usually at the speed of an advancing glacier without the help of global warming. Meanwhile, the most popular activity is to point at somebody else at the table and insist they're not holding up their end.

Blaming others or the team in general for your failures is a losing proposition. Imagine this very imaginable bit of *"Must See TV"*:

> Project Manager: "Mr. Trump, it's true we failed to get the product to market on time, but several team members dropped the ball and there's plenty of blame to go around. If you'll just look at ...."
>
> Mr. Trump: "You're fired!"

As The Donald knows only too clearly, ultimately, the blame falls on the project leader. Here's a tip: If you want to get something done, and you can figure out what's happened in the past that prevented progress:

1. Take care of it yourself;
2. Come up with a "workaround" to bypass a problematic person or issue;
3. Build in some extra time;
4. Work longer hours (and weekends); or
5. Get help from a supplier (that is, a supportive outside party).

I've had to use each of these tools at different times to keep projects moving forward, and even then I've been behind schedule.

But the best – and probably the most difficult – option, is changing the habits of your team, even (especially!) when you're in the middle of a project. You want to positively influence "Groupthink," a phenomenon identified by psychologist Irving Janis in 1972. The term describes a process by which a group can make a bad or irrational decision – or fail to make a good or rational one – due to a certain negative momentum. No one steps up and tells the King his pants are around his ankles; everyone just plays along.

How do you overcome Groupthink? One way is to introduce alternate opinions based on irrefutable data, and do it in a constructive manner. Find a way to lead the team towards a proposal that better positions them for success. Of course, that's easier said than done. Six months into the project, even after constantly presenting schedules with realistic end dates, no one on this team has the *kahones* to confront King Richard.

Part of their rationale no doubt was: If he hasn't realized after almost a year's worth of delays that they're in trouble, talking to him in a clear manner with objective data probably isn't going to change his mind. Of course, that's no way for a project manager to manage a project. It's amazing how, when you start to assume or predict "negatives," your assumptions and predictions come true.

It's one thing to use data, such as a schedule or action item list, to show everyone the current state of a project. It's a completely different animal to overcome a deeply rooted organizational culture. For example, clearly Operations is dealing with a group of very smart engineers who, nonetheless, have little or no concern about the processes and functions that receive their handiwork. They're used to working a certain way – it's ingrained in their psyche. In fact, Dean is amazed that Operations constantly drags him into their world. He says that, in the past, they just handed things off to Production.

There's no way this team can truly be successful if it doesn't work together and eliminate its sub-optimizing old habits. It's not that Operations wants Design to "hold its hand" for the duration, but just long enough for Operations to get the factory running.

Maybe the "old habits" of the engineers were perfectly acceptable on past projects. The problem is: the needs of the present program fall outside their skill sets. Most of them, we might imagine, cut their teeth on high-tech, low-volume government work – whereas this project requires adhering to an industry standard for high volume. It's a huge leap to try and plan, cost out and develop timetables for time- and cost- sensitive products vs. a "technology at any cost" product. This team's problems aren't due only to the project's lack of business processes, but also to a mismatch of skills to project requirements.

This is a dilemma for a project manager, who often doesn't choose team members, and doesn't fully understand the mismatches in the beginning. At the point in this project that our Project Manager figures this out, it's too late to turn back. He has to try and pull out a win with the players on the roster.

This is one of the greatest lessons and it can't be overstated. In addition to defining the end goal and following an established commercialization process, you have to make sure you have the right skills on the team. Fortunately for me, in most of my previous project experiences, I was part of a team that possessed the appropriate skills to get the job done.

At first glance, it looks like our project team's design skills are more than a match for the challenge of its proposed low-band products. But as the football saying goes, after further review – over six difficult months – many of their assumptions turn out to be dead wrong. The managers are throwing everything at this team because the individuals have been successful in the past. But how can these individuals succeed on a dynamic new project with skills better suited to the past? For my next project, I'm going to develop a "skills required" matrix and compare it with team skills in order to identify and address any shortcomings.

Chapter 7

*"Each day we battle the supply chain, constantly gathering
intelligence on parts delivery, and hoping Design won't launch any
new offensives. But we are losing the struggle on several fronts."*

December 4

Dear Diary,

Here's some great news to start the month: We've actually
started building one of the new circuit boards at ACME. Over the
next couple weeks, I expect we'll be in a constant state of flux as
we place other boards into production based on design readiness
and parts availability. This is going to drive our contract
manufacturer crazy, but we warned them, even going as far as
sharing our proposed build schedule.

We've started to plan for upper-level items – enclosures,
hardware, power supplies, packaging material, data plates – in
parallel to the production of new boards, so customer trials can be
built. Sal pushed for delivery dates again and complained about the
problems he's having with end customers due to our ever-changing
schedule. The real problem is he hasn't attended a production
meeting in over a month. I guess he thinks we can magically meet
dates. My respect for him has diminished. There's no way he
should be promising dates without understanding the state of the
project – and, more important – of our track record (or lack
thereof) for meeting a committed date.

To build other boards, we're at the mercy of our supply
chain, which hasn't been given enough time to specify, quote, order
and place parts on the boards. The board queue has been changing

daily as we get a better handle on parts delivery. Overall, the team seems to be confident that we can build all new boards this month, with some work-around for components that we know won't be available. I find this enthusiasm amazing, given the problems we ran into last time – missed parts delivery dates, design changes and limited time to assemble the builds. But, hey, I'd rather see 'em smiling.

And here's some not-so-great news to start the month: Bare boards due at end of this week are being delayed by two-to-three days. The warning came too late to do much about it – and, worse, it's from one of our preferred suppliers. How do you plan for these things? No amount of early "what if" analysis would have helped. We've had to quickly change the board queue (again!) and make the most of a bad situation.

Each day has been a battle with the supply chain, constantly checking on parts delivery and hoping Design will not request any new parts (wishful thinking). New parts requests have been coming in and new parts orders have not. Peter was off again trying to find a supplier for a handful of resistors – without them, the project could come to a screeching halt. On the other hand, Design is testing new boards and initial results look promising.

This project has been severely understaffed from the start and things aren't getting any better on that front. A lack of sufficient design technicians, a Quality group and a production facility caused us to approach the next phase at a snail's pace. The promise to deliver customer trials at the end of the month isn't going to be kept (I promise). This will send Sal through the roof. Why do we over-commit, given the breadth and depth of the tasks staring us in the face? We're really all guilty of pushing on the limited resource bubble. In our zest to get new boards to

Engineering, we shortchanged the time needed to produce the final product.

December 12

Dear Diary,

We've moved into the next phase of the project (sort of): Figuring out how we're going to build and test units. A few of us in Operations visited Engineering to go over how they're currently producing engineering models (EMs). Somewhere in the midst of all of the craziness, I took on the additional job of developing a rough manufacturing plan. The overall build isn't complicated – at least at first glance. How hard can it be to assemble a handful of circuit boards into an enclosure?

Famous last words. The industry standard for product certification requires testing each unit at various temperatures. Even though overall labor content is minimal, throughput time is doubled due to temperature testing. We're going to have to figure out a way to improve throughput without spending a significant amount of capital. Too bad we don't have a business case that could provide some guidance on capital. At least we'd have some pre-written approval. Battling Richard for capital is going to be agony – just like it was for pre-production purchases.

I was going over labor hours required to build units and trying to determine how much time we'll save with learning curve effects – and it dawned on me: Sales hasn't provided volume. Richard has given us some indication of anticipated demand, but each time we've met he's thrown out drastically different numbers. Again, having a business case to help drive this program and guide these decisions would be wonderful. It provides a good sanity check and do we ever need our sanity checked from time to time.

December 23

Dear Diary,

A majority of the team– including me – is taking vacation this coming week. Good timing. Everybody needs to recharge and prepare for the post-holiday chaos that's sure to visit us. It'll be great to be home with my family, even though the week usually flies by. Maybe over the holidays the Process Fairy will sprinkle Process Fairy dust over our team, so we can come up with a more realistic schedule that drives units into the hands of our wildly satisfied customers without delays. Yeah, I know, I've got visions of Process plums dancing in my head. A guy can dream, can't he?

We just had our last team meeting and it was a battle royal. Sal was at his wits end about us missing customer trial dates again, and I don't blame him. But no one on our team has had a heart-to-heart with Richard about the reality of the project, especially the difficulty in designing a complicated project to a new standard under such extreme time constraints – not to mention the fantasy of his demands. Who wants to step into that buzz saw? Okay, maybe I'll volunteer (again).

The lead engineer, Will Invent, expressed confidence that product can be delivered during the first week of January if he receives one of the new boards by the end of month. And, yes, he knows we'll be out for the holidays. Bless his heart. Will is a great guy, but Operations has figured him out. He over-commits and under-delivers. Translation: Add at least one week slippage factor to his end date, and another week for funsies. Will is the perfect political animal in his ability to give the appearance that he can save the day. But we all know about politicians.

Back in the real world, Operations decided to get a head start on labor costs for the product portfolio by gathering quotes from two key contract manufacturers. The released product BOM, mechanical drawings, parts specifications and volume estimates (that is to say, volume guesses per Richard) were sent as part of the quote package. The goal: have cost estimates back in our hands by the first part of January. The team realized that unit-manufacturing costs were above initial goals and wanted to get a better handle on overall labor dollars. This will give us a good sense of how much more work we have to do in order to reduce product cost. Call it a Christmas present to ourselves.

December 28

Dear Diary,

And we're back. The week, or the portion of it I managed to stay home, definitely flew by. Actually, quite a few of us couldn't stay away. I just talked to Will, who mentioned that one of the new boards isn't running consistently. It'll probably need to be redesigned and thereby cause a delay of three weeks. Here we go again. No formal plan in place that clearly defines what's required to design, test, manufacturer and sell product – so the project, once again, is taking on a life of its own.

To add insult to injury, Purchasing reported that the last seven parts needed to complete the remaining two boards are late. This means we won't have all of the new boards built in December like we wanted. What a way to end the year. Even though our team has made progress, you're only as good as your last project update. This one is not so good.

The more we've changed the design and added new parts, the more uncertainty we've generated in the supply chain. Operations – and no one else – committed to having all of the boards built by the end of December. We've got well over one-hundred years of experience among us, but we failed to accurately predict an end date for the board build. And more importantly, we didn't take into consideration any "what ifs." (Like: What if we don't do a "what if" exercise?)

I hate to miss a commitment. True, project management is no picnic, and nobody remembered to bring their crystal ball to this party. Even with this group of bright minds, relying on best-in-class suppliers; we still fell short. Clearly, simply padding each task to plan for the unexpected is unacceptable. Groupthink takes over and the project falls in line with this bloated schedule. What I need is a conscientious plan to steer Groupthink to the project's advantage.

Hey! That's good! I never thought of that before. Now the question becomes: Just how exactly can that be done? I hope there's some of that holiday cheer leftover.

Learning #7: Parts

Parts is Parts

Parts is parts. Like death, taxes and gravity, there's no getting around it. This has, and always will, remain true – forever. It takes time and effort to research, specify, quote and order parts. What teams usually don't plan for is a part's lead time. There are many ways to attack lead time, but you have to understand the standard before you improve it: Order your parts sooner, get to market sooner. Simple, eh? But that equation is complicated by this question: How much risk are you willing to take?

Richard is "risk averse" in the sense that he spends a lot of money and energy on R&D, premium-priced build and useless boards. Of course, you want to make sure your design and assembly meet or exceed the customer's expectations. But there are ways to manage that process to optimize efficiency. For example, if Richard had allowed his team to pre-buy some of the longer-lead parts, he could have made more revenue sooner.

Unfortunately, design engineers aren't supply chain professionals, and they're not paid to specify time frames for the components and materials they recommend. Their job is to create a functional, quality product. What's required is someone that understands supply chain and can work early on with the engineers on the team to address parts issues up front. He or she should understand parts alternatives, parts cost, parts suppliers and how to plan for them. Good supply chain experts attack lead times to enhance time-to-market and increase the responsiveness of the factory in meeting customer demand.

It's imperative early on in a project to include someone with expertise in supply chain to steer the team in the right direction. On this team, Peter has the expertise, but Richard is effectively neutralizing him and others in the organization. Most of the successful projects I've been a part of recognized the importance of getting a jump on the supply chain. We were able to match demand with inventory as soon as final approval from QA came in for the product. We didn't have to wait for the supply chain to warm up. In this case, warm-up based on long-lead parts is eating up almost four months! Fortunately, Peter is mitigating these lead-times by buying material without Richard's approval.

World-class supply chains start at the top. Isn't that what every Harvard Business Case states? A world-class company invariably gives credit to some visionary that led them to greatness – Jack Welch at GE, Andy Grove at Intel, Bill Gates at Microsoft, and so on. Well, Richard isn't in this category. In his view supply chain management is, at best, a necessary evil that impedes design.

Even with the team pushing constantly to start the procurement effort, and producing schedules showing the criticality of such an effort, Richard hesitates for weeks. Bottom line: It's extremely difficult to get the supply chain running efficiently – so why leave it until the last minute?

Chapter 8

"After months in this engineering-centric, strategy-starved working environment, even Operations is losing its edge. It's a cultural tsunami, and darned if I haven't misplaced my surfboard."

January 3

Dear Diary,

Well, I'm still here. I guess Santa had me on his "naughty" list. Anyway, Happy New Year! Last thing I remember before all the eggnog, we'd missed our goal of building all the new boards by year's end. OK, not good, but we did make a fair amount of progress: The new board assemblies have been released into our MRP system and product BOMs have been structured. We'll get more good news when the last two boards are built and shipped (to three separate design locations … ahh, efficiency).

We're a couple weeks late against our commitment. But given the holidays, the new design direction we've taken and the hundreds of new parts that had to be procured, this isn't all that bad a position to be in, I suppose. My New Year's resolution is to not have anything go haywire anymore – and world peace.

While the final boards are being produced, Operations has been trying to figure out the investment needed to produce in volume. What volume? Heck, we don't know. I developed a spreadsheet – based on both identifiable and "guesstimated" figures – that includes initial costs for capital, training and facility modifications, plus the ongoing building and testing of products.

And Peter generated an MRP report that shows costs for our next material buy. This at least gives us a starting point.

Richard never makes a decision without financial data, and if we're going to maintain our momentum from the last two months, we need his approval to set up the factory. Low and behold, Richard made a trip this past week to see the Design team, and decided to stop by to discuss start-up costs with us. Meetings with Richard haven't gotten any less painful. I doubted this one would be any different, remembering that old habits are hard to break.

Operations shared its cost data with Richard, and it became obvious to everyone present (including Richard) that without major cost reductions, the product won't be profitable. Richard decided that the engineers would have to redesign it to eliminate a handful of very expensive components. To the team's surprise, he also provided board assembly and labor cost targets to get us back on track. Where these came from, nobody knew. It wasn't the first time Richard presented us with information that was untraceable except to the recesses of his own mind.

Trying to achieve these targets without performing a "witness" assembly on working (or any type of) units frankly scares the hell out of me. I think we're being setup for failure, because the targets aren't based on anything tangible. We'll see. Operations volunteered to lead the cost-reduction effort, in part to be able to maintain some control, but Richard took ownership. To quote our Fearless Leader: "Engineers think only of functionality to specification, not about cost and/or profitability." Not very poetic, but no one disagreed with his statement.

January 11

Dear Diary,

I never could keep a New Year's resolution for very long. Ongoing problems with the new digital board have things at an impasse. (I hear world peace isn't going too well, either.) The boards were delivered weeks ago, but our progress is imperceptible. During the Wednesday call, Richard and a few others suggested trouble-shooting ideas. Design is open to anything that can help break the bottleneck.

Operation's is currently analyzing the product and factory set-up costs to see where cost reduction efforts should be focused. This is shooting in the dark (no we're not in the porn industry), since a working unit is a ways off and we could be spending time on a part or assembly that won't be used. We did a quick Pareto to identify the top ten items to tackle, and shared it with the engineers to see if these will be part of the final design.

Then we decided to put together yet another spreadsheet with cost-reduction ideas for Richard. If we can't own the process, maybe we can influence it by "using" Richard to push Engineering and others toward a lower unit manufacturing cost. It's not that he won't take the lead – he said he would – but we can't wait on his timetable. Some ideas affect parts, which have lead times, which depend on engineers using these parts. Peter has shared the next round of build costs with Richard. We know he's going to ask for more cost-reduction ideas, with or without a meeting.

Over the years I've learned that a big part of project management is getting to know the players involved. Operations is getting a better handle on how to deal with Richard. Hopefully this will save us some time. The spreadsheets we're preparing aren't a bad way to communicate. But they take time (that's why we're trying to save some) and our team is overloaded. Team meetings resemble hunting expeditions as members fumble from spreadsheet to spreadsheet trying to find the correct information.

January 16

Dear Diary,

It's mid-January and we haven't received a quote yet from either of the contract manufacturers. We really need this information so we know where to go once we settle on a design. In the meantime, a group of us has been working on labor standards for major assemblies, required tests and final packaging for the product. Wouldn't it be great if we had a previous version of our product, with a predicted learning curve, so we could actually base estimates on facts? But to put it bluntly, our revised production labor standards are WAGs; that is, Wild Ass Guesses. As with previous aspects of the project, decisions are being made without data – a dangerous thing to say the least.

Then there's the issue of morale. After months of slogging through this seat-of-the-pants, engineering-centric, strategy-devoid environment, even Operations is falling victim. It's just too hard to continually swim against this cultural tsunami. As the project manager, I have to rise above this and keep pushing our team. But the way things tend to be handled around here, plus our limited resources and this mountainous task we've got before us, it feels more like pushing a rope.

Design has given up on the current digital board design, so Dean puts the majority of Engineering's resources on two parallel redesign efforts. Basically, two different teams of engineers are racing to design a working digital board. It's a race to see which

80

revised board will be given to Manufacturing for fabrication (The sooner, the better). After all these long painful months of work, the product still isn't functioning. Customer backlash from continually missing due dates is escalating. And the sky is turning blood red (well, I made that last one up).

People always say they need more money, more hands on the oars, more time. But it's become clear to me, given the amount of work we have to do, the staffing levels and resources we can assign against it are inadequate even in a perfect world. Designing, testing, manufacturing and shipping our technically advanced product in the time frame Richard, Sal and our customers expect, is now well beyond unrealistic. Our project's faucet leaks from more than a few joints. Results are coming very slowly – certainly more slowly than any of the managers think is acceptable. But even that stark reality doesn't seem to be enough to force a significant change in the way we do things.

January 25

Dear Diary,

Then again, maybe it is. Richard finally addressed the issue of progress – or lack of same – at our last team meeting, after hearing that even the software team has fallen victim to delays to the tune of three weeks. He asked for ways to speed time-to-market after kindly questioning why we can't seem to meet a committed due date. There was a long silence before he broke the ice and started to brainstorm ways to get us back on track – which to me was a pretty good (if tardy) way to handle matters at that point. Suggestions included hiring more people, working on just one product at a time, and quickly turning around the redesign of the digital board.

Kudos to Richard for that, but it's about time he stepped up. The truth is, the only thing that can really be done in the immediate time frame is to throw resources at the digital board. Hiring new people was mentioned months ago, but the idea was nixed because everyone thought it would take too much time to get a "newbie" up to speed. In retrospect, we should have hired at least two more engineers and technicians back then. Why can't people be smart in the beginning, before problems occur?

Even so, it's not like Manufacturing has been sitting idle. We created a Sales & Operating Plan, complete with projected volumes, to determine the investment required for the next build. This figure is really what Richard is willing to risk, not any long-term projection. The supply chain will be engaged to address this subset of overall demand, and then be abruptly shut down again. With all the issues we've had with part lead times, starting and stopping the supply chain is really going to seal our fate regarding meeting – or not meeting – customer dates.

Meanwhile, Operations has done some work with ACME Inc. to set up a potential schedule for executing the next revision of the board. But as the month ends, quotes from our potential contract manufacturers still haven't been received. This situation is doing absolutely nothing to pull me out of my post-holiday funk.

Learning #8: People

Leader

As much as the term "leader" is overused, every team needs one. A team's success is usually proportional to its leader's ability. I think of a successful leader as part Drill Sergeant, part Personal Trainer, part Mentor, and part P.I.T.A. (no, not the bread, or the animal-friendly organization … think about it).

An effective leader is someone who pushes everyone to a common goal that is difficult to achieve. This individual manages to get more out of people than they thought possible, and gains their utmost commitment to the goal at hand. A leader is someone who makes the "whole" greater than the "sum of its parts." He or she is adept at cultivating or, if need be, overcoming the corporate or departmental culture. And of course, these types of people are rare.

There are countless books, seminars, courses and live presentations on leadership that, at least in part, attempt to uncover and develop the next Jack Welch, the legendary former CEO of General Electric and the author of several excellent books on leadership. But identifying, then making effective use of, this sought-after species remains a challenge for business. Companies tend to do a poor job of spotting and recruiting leaders, and end up putting individuals in positions of leadership who lack the necessary skills.

Richard, for example, would never be confused with a great (or even a good) leader. He's a brilliant engineer, but he lacks the ability to "manage" his resources – to line things up and get a common commitment – as well as complementary, if not coordinated, action – to commercialize his products. He constantly

83

oversimplifies things that he doesn't truly understand. These gaps in understanding are mostly related to managing production – from purchasing of materials through final tests and packaging. So he ends up handcuffing the experts he hired to run the factory.

As we learned earlier, people have a tendency to hunker down in their own functional foxholes, and they don't peek their heads out unless absolutely necessary. Each function has people who are very busy doing the particular thing they do best – but that focus can be to the detriment of the overall project and ultimately the customers. This is what W. Edwards Deming referred to as "sub-optimizing the system," which undermines the idea that the whole is greater than the sum of its parts.

It's the effective leader who can get the varied people on his team to peek out of their silos, and to do it more often. Instead of working diligently within a particular silo, team members must work as a team with a common goal. Regarding our project, management, especially Richard, is responsible for articulating the project goal, and for aligning team members toward achieving that goal. There are various leaders at various levels involved in any project, but there must be a Supreme Master who keeps everyone focused on the overarching goal so they don't return their gaze to their function goal. That's something this team lacks. So the project struggles. It's not all Richard's fault, but a lot of it is.

As the project manager, my job is to ensure that we are on track to achieving our goal; however, this is made impossible because we do not have one goal, but separate ones per function. Since operation's depends on these functions to deliver the product, the common goal ends up being dropped in our lap in the form of a working unit. The problem with our current approach is the immense waste in resources, time, dollars and lost revenue the organization has endured.

Someone, anyone, has to step up and lead our team. The leadership must come from one of the four vice presidents or CEO; however, Vice Presidents have had their decision rights neutered over the years by CEO. Therefore, they do not want to risk their artificial balls. Without leadership, we will continue to struggle. I've tried to bring some discipline and leadership to team, but roadblocks in the form of my manager, Richard, and CEO have totaled my vehicle – I guess it is time to buy another vehicle and jump back into project. Good project managers must have an abundance of tenacity, optimism, and humor. Time spent complaining about problems is time that could be spent fixing them!

Chapter 9

"The CEO doesn't owe us a salary if we don't deliver a working and profitable product. Still, some of the prima donnas on this team walk around like Paris Hilton, spending Daddy's money without a care in the world."

February 3

Dear Diary,

Apparently in honor of Ground Hog Day (let's just assume that's the reason), Design finally provided documentation for the redesign of the critical digital board – and Operations got to work. We've started quoting and ordering bare boards on a fast turnaround, while at the same time drawings are being sent to ACME Inc. to prepare for production set-up. I began to hope that fuzzy critter saw its shadow and that it meant six more weeks of this kind of team energy and effectiveness.

Not. Within hours our vendor discovered numerous errors with our BOM – missing parts, duplicate parts, parts with the wrong pad geometry. We just can't afford any more delays, and yet it's our own internal documents causing the build to be a couple days late. We quickly resolved the board documentation issues and worked with the manufacturer to minimize further delays. I'm trying to remember that the goal was to have these boards built and in the hands of Engineering two to three days after all parts arrived. This board is so critical that Richard and our COO, Sarbenes Oxley, called Peter three days in a row to find out its status.

So now the team is in a holding pattern, anxiously awaiting the arrival of the board from ACME. It will be picked up at the vendor's, hand-carried to the airport and sent to our engineers. I've seen a lot of projects develop bottlenecks that force the majority of the team to put on the brakes until the accident is cleared from the highway. Now I've seen one more. Most everyone on the team has been rubber-necking this tragedy and praying that things will work out.

What puzzles me is that weeks, actually months, have passed since the first revision of the digital board was put into the hands of Engineering, with no sense of urgency. But now we're bleeding from our ears trying to save hours. What's not puzzling to me is that, no matter how fast team members drive their individual tasks, our forward progress is going to continue to be limited by bottlenecks.

February 11

Dear Diary,

Good news: One of the team's boards is finally in the hands of our engineers after a world-class effort by Operations and ACME Inc. It's just before the weekend, so the results of the redesign won't be known until early next week. Of course, hardly anyone around here works on the weekend, which drives me crazy. Instead, the engineers will arrive on Monday to analyze the boards. Richard won't put down the hammer and get them to work Saturday, Sunday or both. Our corporate culture – actually, a lot of corporate cultures – lack a sense of urgency. Some days it feels more like a government agency around here, and that's one big reason why we're so far behind schedule.

In Operations, we're preparing to build the other version of the digital board from the second engineering team. The race continues to see which team has the winning design – which is the one that will best help get us out of our current predicament.

More good news: Design says the initial digital boards from ACME are "working," so now the team is working feverishly to integrate that board into the product. There's still much to do, but for now we have some success to build on as we head toward the final product design. I expect that attention will turn to Operations more and more to get parts ordered, boards built and a factory set up – preferably at breakneck speed.

Our next step is to review quotes from the potential contact manufacturers, and then ask Richard's permission to order the next round of parts. Quotes have been coming in and – ouch! – The current design is far above our target cost. No real surprise to Operations or Richard. We've known for a while that this product needs a major cost reduction. The disappointment is that we haven't made any real progress on this front.

Not good news: Our workload keeps growing while our resources remain the same. Now we have to build new boards, plan for the factory, reduce product costs and tackle the RoHS directive, all at the same time. I can see that trying to manage all of this is going to be a huge headache, but others in Operations and our sister company will step in to help (I hope).

PM Lesson #2,827-A (or whatever it is): Being a good project manager means knowing when you need help. And ladies and gentlemen, I need help. PM Lesson #2,827-B (I presume): Being a good project manager means not being afraid to ask for that help. I can do that too. Don't want to struggle through these activities solo only to be marginally successful. Right now my job is to get the low-band and then the high-band products into production, and we need everyone to be focused and to pitch in. That's my pitch to the team members, in group settings and in one-on-ones.

Fortunately, we've hired some engineering co-ops to help us with RoHS. Isn't it nice to have extra people at your disposal? I better load up their schedules before somebody else does.

February 19

Dear Diary,

The second engineering team's version of the digital board will be built over the next couple of days. Dean and Will are chewing off their fingernails waiting for this board so they can determine which design survives and will be used in the final product. Operations recently met with Richard to map out product priorities, parts procurement approval, testing assistance and UMC cost-reduction guidance; and also to discuss pre-production strategy. It was downright frustrating. Richard (as usual) was hesitant to take even the slightest risk, even though we're facing a critical decision.

Bottom line: If we don't order parts now, this program is D-E-A-D! Information trickling in from the marketplace tells us that some of our competitors have moved ahead in the race and we've become spectators – because of Richard. There's no possible way, given lead times, to build these products in time to appease our already irate customers, not to mention to supply units to the certification facilities.

But I don't want to throw in the towel just yet. To help push the decision, I presented an updated schedule with activities required to deliver the product in June and to showcase the criticality of ordering parts. After much discussion (where my passions were running pretty high), I got approval for Operations to purchase the next round of parts. As part of this new push, we formed two teams to address cost reduction and testing strategies.

I know, I know. These are items that should have been defined at the very beginning of the project, not at the 11th hour. March is right around the corner and the schedule shows that every board in the portfolio requires revisions. The expectation is that these final modifications will lead to a sellable and fully functional product. Ah, but haven't I heard this before?

We're back in the movie *Groundhog Day*, where we relive what's happened before, but now the plot centers on the new modular design. Changing our design strategy didn't magically yield a whole new corporate culture – or dampen the consistent negative vibes from our customers. What unsettles me is that hardly anyone seems to notice or care. It's become white noise that's easy to ignore. The lack of attention and dedication to the customer, the folks who pay our salaries, is ridiculous (not to mention myopic).

Even the news that our competitors are in the lead in terms of development is falling on deaf ears. The CEO doesn't owe us a salary if we don't deliver a working and profitable product to customers. Still some folks on the team are walking around like Paris Hilton, spending Daddy's money without a care in the world. Having worked for a company where the threat of layoffs was a normal part of life, I know how damaging this attitude can be. I find it just unacceptable.

We're eight months into this project. If we don't get a grip soon – develop a sense of urgency, face some hard decisions and make some hard choices – we can toss this whole project in the trash can. I'm really counting on the cost-reduction and testing strategies teams to "put some meat on the table" and force our hand to act.

February 24

Dear Diary,

The second version of the digital board was assembled earlier than expected and units were shipped directly to our engineers. Once again, ACME saved our ass. The team has had a series of wins over the last two weeks – not rearranging deck chairs on the Titanic, real progress. But there are more battles to fight. The last weekly meeting got pretty hot thanks to a discussion of early customer requirements. Richard, Dean and Sal all shared different quantities and delivery dates for customers. Along with the anger, we were all treated to a heaping helping of confusion during the call.

At this point, all Operations wants is an agreed upon list of what's feasible and when. After some (more) verbal warfare during the call, the team finally agreed to dates that we hope are, and will remain, feasible. But this inability to communicate has caused some permanent scarring of team members – not unexpected given the time pressures, but not a good thing.

The Forming – Storming – Norming – Performing model of team development, first proposed by Bruce Tuckman in 1965, applies to our team. But we've regressed back to Storming after months of hovering between Norming and Performing. I'm not sure if we can overcome this and get back on track.

Like our previous schedules, due dates for the new board revisions have already been trashed. Fortunately, these boards are not on a critical path (yet). But when is this team going to get things right? Even when I suggested adding additional time to tasks because we're understaffed, no one agreed. We're putting an honest assessment of our activities on the shelf and only showing best-case scenarios that we haven't been able to achieve for the better part of a year.

A lot of people on the team are just in a state of denial. I think they're afraid to disrupt the status quo because having a lousy job is better than having no job. Result: later and later launch dates, angrier customers, large budget overruns and a tarnished team image. This is the kind of month where you just want to go home and hug your kids.

Learning #9: People

Communications

Next to a great leader, good communication should be an obvious requirement for any successful project. But even in today's wired world, where e-mail communication can be sent anywhere at any time to anybody, problems involving communication still plague projects. That's because we sometimes equate advances in communication technologies and techniques with improvements in human communication skills. They're not the same thing.

For example, it isn't unusual for an e-mail to be forwarded to many individuals with no one taking ownership. The problem occurs when task due dates pass, the project leader asks for a status and team members deflect responsibility – rationalizing about how they thought someone else who was copied in the e-mail was working on the issue.

Teams must plan for communication with as much diligence as project tasks and delivery dates. The team in our story has quite a few meetings, but they're conference calls. Meeting effectiveness would go up exponentially if some of these gatherings were face-to-face. There's nothing better than being able to read someone's body language to understand how well a communication is received and retained – and, more importantly, whether there's true commitment to a particular issue or action item. Operations has resorted to daily meetings to track progress on key milestones and help drive some progress.

Peter and some of our other characters have suggested periodic face-to-face meetings to address critical paths and get answers to critical questions like volume, units needed for regulatory approval, strategy for RoHS/WEEE directives, channels, service plans, training, and dates for key customer trials. But no one seems to want to do what it will take to improve the current situation. Along with an unwillingness to commit the resources, a functional culture that isolates team members in their own particular worlds is hurting any chance of working more effectively as a team. Everyone continues along on their painfully slow paths toward a product launch – assuming, of course, that there's still going to be a product launch.

As we learned earlier, the first thing that should have been communicated to this team was a project strategy and an organizational direction – a simple mantra that's clearly understood with no room for misinterpretation. Like this: "We're going to produce low-band products for this customer group – in this volume, at this price, with this target cost, through these channels, over the next twelve months." Clear enough?

However, most any kind of communication at most any point in the project is better than no communication at all. It's like Jell-O: There's always room for communication! A schedule and an action items list are part of any good communication mix, but much more is needed from leadership to apply pressure on the right priorities. As we can see, the lack of an overarching message from Richard allows each person to filter in or out what he or she wants and come up with his or her own idea of what needs to be done. This behavior though commonplace in many companies, is very detrimental to the project and more importantly time-to-market.

Chapter 10

"Everybody knows that waiting for unexpected problems or obstacles to magically improve on their own is just plain stupid."

March 3

Dear Diary,

We're at the tipping point: March is going to either make or break this project. This is the month that will see a perfect storm of success or disaster. I really don't see a middle ground. Revised boards, software and hardware are supposed to come together into what will be our first functioning engineering models (even though Richard and Sal would like to call them pre-production units). This is also the month that we promised delivery to key customers.

To meet these new stake-in-the-ground commitments, some of the engineers are working on trial units. The downside is that it puts us in Project Purgatory: The product isn't stable enough to transition to a contract manufacturer, so the engineers have to build early customer units instead of concentrating on finishing up the design. I'd like nothing better than to transition the product to our CM, but a lot of work remains before that can happen. I trace it back to never hiring the extra technicians and engineers that could've helped at this stage. If we had our own operators, some could've been assigned to design tasks earlier in project to spare the rest from building units.

Meanwhile, at our team meeting we saw signs of progress on board design. Design had the nerve to ask when production would be ready, since they can't continue to support trial build and meet the project schedule. It's a Catch-22: Operations needs a product to build but we have no product. How Design can be oblivious to this fact is incredible. All we can do is continue to push for final board revisions. At the time of the meeting six out of six board revisions were late per Dean's schedule. Everyone is aware of this because it's tracked as part of our team meetings – which now occur every single day of the week.

There's still a lot of work to get done before product launch, so Operations set up a meeting with Richard to get approval to move forward with the CM. We calculated cash outlay for the initial product build. Our goal was to try and alleviate the constraints on project resources and prepare for long-term production. As usual, Richard questioned the numbers we have for final product cost before we ever got to the overall budget. He's known all along how far off we are from cost targets. He's supposedly leading the cost-reduction efforts. And less than two months later, he says he's shocked – shocked! – by where we stand. The term my college students use for people like Richard is "tool" – or more explicitly "What a tool!" Hopefully this doesn't need further explanation.

Result: We have to provide even more detailed cost data to him before the end of the week. As John Stossel says, "Give me a break!" I thought Isai was the biggest human impediment, but Richard is in a whole other galaxy. No matter how hard we prepare for these meetings to drive fact-based decisions, Richard is just not capable of making an expeditious decision – even with mountains of data to support the appropriate ones (and provide political cover).

March 10

Dear Diary,

Operation's is extremely frustrated. Why? Because we can see the future, and it is dim. All board revisions are late, Dick (that's what I'm calling him from here on out) won't give us the go-ahead with the contract manufacturer, part due dates are slipping, product costs are – surprise! – over target, and we're unofficially, informally being leaned on to make-up for everyone else's shortcomings.

Operations also can see what will happen next: All board revisions will be released at the same time, meaning that Operations will have to scramble to manage another unworkable situation. As a courtesy, Peter and I give ACME a call to tell them (warn them is more like it) about our prognostications. They think we're joking. We're not, at which point it got real quiet at the other end of the line.

It turned out that four of the six late board revisions were released within days of our call to ACME – so we're not going to open an Operations Psychic Hot Line to predict lotto numbers. We called ACME immediately to discuss the build schedule – only to be told that they're buried in other work. Our expedited build of two to three days could take weeks now – not the news we wanted to hear. But these are common consequences of blowing up your own schedule many times the consequences of doing so are very painful.

We agreed to give ACME time to figure out the best way to meet our needs as well as those of its other customers. One of the things I never asked was whether they had seasonality in customer demand. Now I have my answer: ACME's busiest month is March, and the (disastrous) perfect storm is gathering strength.

Dick recently showed up in town to address employees on yearly achievements. Our team took the opportunity to set up one more meeting to discuss product costs and the next steps for production. Dick provided an updated list of target costs for the board build and final assembly, and we agree to pursue these with suppliers. His updated numbers were in his head (or up his – never mind). Dick rarely puts anything on paper so it can be broadly communicated.

Peter and I decided to set up a quick trip to visit with our design engineers, a potential contract manufacturer and ACME. Our objectives were to perform a witness assembly on products, and address target cost goals and due dates for boards. Back soon.

March 18

Dear Diary,

On the first day of our trip, Peter and I met with our engineering team and spent the morning performing a witness assembly. Documentation is lacking in many areas for the build and BOMs are inaccurate. But it was great to actually spend "face time" with those folks and review other aspects of the program that need to be addressed.

The next leg in our journey took us to ACME to negotiate price and determine the delivery date for three of the four recently released boards. We arrived and were welcomed with open arms. They've been wonderful partners so far, helping us compensate for delays by turning boards in two to three days when they could. This is the same lead time we need now to keep us moving forward. They said they would try.

Then the wheels came off. On the last day of our visit we got hit with bad news on cost targets and board due dates – ACME's best-case cost reduction is 10% vs. a goal of 40%. Also, turnaround on boards is going to be two weeks. Before heading home, we sent an e-mail to everybody on the team sharing results of our visit and explaining the reasons why. I can't wait to read the responses. This project has gone to the "quick turnaround well" too many times. At the moment, the well is dry.

On the bright side, ACME always seems to find ways to improve their delivery dates. I trace this ability back to the leader of the production team. Peter and I know she'll do everything in her power to help us out – with the same passion I'm sure she provides to all of ACME's customers. I've been fortunate enough to occasionally work with world-class suppliers that acted as an extension of our team. ACME Inc. is that kind of a supplier, and with them we have that same partnership.

In fact, our partnership with ACME is better than the relationship Operations has with many of our own team members. How sad is that? Maybe we can hire Production Lady away from ACME.

March 26

Dear Diary,

We returned from our trip very disappointed. Operations immediately started preparing to meet with Dick once again about approval to move forward with a contract manufacturer. (How many times have I written that phrase?) We have to make a decision this month, or there's absolutely no chance this project will be even remotely successful.

So we met and, as has been the case all along, Dick nitpicked our calculations because the estimates we have for costs are still outside his goals. (Persistent things, facts.) How does he expect a design that hasn't changed in months to somehow magically be cheaper to develop? In fact, many of the tweaks actually add costs to the product in the interests of making sure it works.

The ticking of the project clock sounds more like an armor-plated Army of Destruction marching down on us. Yet again, a decision to move forward with a contract manufacturer has been delayed. We've now spent well over a month trying to get a decision. No amount of pleading on my − or on anyone else's − part has convinced Dick to pull the trigger. This is the part of a project I hate the most − knowing what needs to be done, and being kept from doing it for no good reason. Basically having your decision rights neutered!

There's a team of people brainstorming ways to save labor costs and still get production started at We Build Stuff (WBS), a contract manufacturer near our engineers. This is a good thing. Since the design is still in process, proximity allows any issues that appear at WBS to be quickly addressed. The downside is that the engineering team gets diverted to troubleshooting. The more issues that develop, the less time they spend on design.

Another issue with partnering with WBS has to do with our documentation. The better the documentation is, the easier and more efficient a transition. So it follows that since our documentation is far from perfect, this transition is going to be turbulent. But at this stage, we have no choice. To save money, we investigated using contract labor instead of WBS employees, and pay the owner of company to manage these folks. This would give WBS and us some flexibility, given the longer-term unknowns related to our product. We shared this idea with WBS's CEO and asked him to get back to us in a couple of days.

Just as March was about to go out like a lamb, ACME delivered a partial shipment of critical boards – much to the delight of the team. (Like I said, it pays to have great partners!) But a decision to move forward with a contract manufacturer still hasn't been made. After all these months, our Management Team has finally concluded that the business case for some of the products is negative. Gee, thanks. Since we never really defined the "End" in terms of the 4Ps of marketing (product, price, place and promotion), and never had a real business case, forgive me if I'm not surprised by this. In fact operations have shared concern with management about negative product margin for months with no one really listening. Now at the eleventh hour management has awakened to this fact.

Meanwhile, Operations has been in a holding pattern waiting for a decision – any decision – to be made. I'm pretty sure this program is heading toward the abyss. Lack of a strategy and Dick's indecision are killing our chances for success. This obvious fact somehow bypassed (until recently) the managers we have in charge. It'll be interesting to see the pressure placed on Operations as the calendar turns to April 1st – Fool's Day – how fitting

Learning #10: Phenomena

Step Back to Move Forward

Being confronted with unexpected circumstances or obstacles, but then just waiting and hoping that things will magically improve on their own, is just plain stupid. Couple this with the inability or unwillingness to make decisions and you impede progress. You frustrate your team while doing nothing to get closer to fulfilling customer needs. What's required is a conscious decision to step back, review the situation with data, and then brainstorm a way to move forward. There's no shame in admitting that you're lost. In fact, stopping and asking for directions is a good thing. At the risk of jeopardizing my Manly Man's Club membership, I confess that I do ask for directions. Plain and simple, our team needs someone to step up and stop the train. Operations has made its piecemeal attempts, but no one, it seems, is willing to go toe-to-toe with Richard.

Situations like this – of the "Oh, shit!" or "This spacecraft will self-destruct in 6 seconds" variety – cry out for classic Risk Management. But at this stage, teams that are behind schedule (not to mention the 8 Ball) usually don't want to fall further behind. So they resist taking action based on this kind of contingency planning, which only exacerbates the situation and generates more opportunities for delays. The Operations team in our story faces a workload that exceeds its capacity, so they're going with the "Peanut Butter Approach": Spread a little effort across a lot of tasks. The problem was, nothing was getting done efficiently and both product portfolios were behind schedule.

If this team, at the early stages of the project, had applied any kind of commercialization process, complete with Phases and Gates, the train wreck would have been predictable earlier on. This would have forced a change in direction in order to get back on track. Or it would have led to an informed decision to cancel the program. That's painful, true, but sometimes it's the only logical thing to do.

These reviews also provide the Gate Keepers (management) an opportunity to adjust strategy based on market and team dynamics. Instead, our team keeps plowing ahead not really knowing where it's headed, and now we are in big trouble with our customers. On top of that, all the delays have put competitors ahead in the race.

Are We There Yet, Diary of a Project Manager

Chapter 11

"No matter what happens during a project, the project manager must never lose sight of the end goal – even if they have to create one."

April 3

Dear Diary,

Rumors at my current company, unlike at my former company, are rare – but we just heard a doozie: Sal is leaving. Most of us brushed it off as a cruel April Fool's joke and went about our business. We're concentrating on when more boards will be delivered from ACME Inc., and when WBS will get back to us about our contract labor proposal.

ACME called and more boards are on the way to the engineers – boy, are they good. We waited for the team meeting to see if the rumor about Sal is true. He lost the respect of our project team since he started isolating himself from us for weeks on end. It's too bad because Sal certainly has a lot that he can provide. But missing meetings at the center of project communications, then promising unrealistic dates to customers – that's just unacceptable.

The meeting got going and we anxiously waited for everyone to dial in from their locations all over the world. Sometimes people are late to call, but not this week. Well, no sooner did we start when Will confirms that Sal has left – in fact, his office is already empty. Since Sal hasn't been involved in such a long time, most people took the news in stride. Some of us immediately started trying to figure out who on the team could go to customers to understand 1) if they still want to do business with us, and 2) when they're expecting their units.

Noll Itall, one of our sales representatives, volunteered to contact customers and told us he would get back to us early next week. I'd call this a good example of a project team that adapts to change at a moment's notice. Particularly in our case, there's no time to fester.

I took the opportunity of the well-attended meeting, to force some more discipline on the team now that we have product catalog numbers and BOMs loaded into our MRP system. (No one has ever accused me of being a wallflower.) Before Operations takes any action, all customer orders must be entered into our system with a due date. This will let us track sales, book revenue, decrement inventory and better plan production – whenever we get the go-ahead from Dick, that is.

The other benefit is that we can show Dick an "open orders" report, which will give him an idea of what customers are expecting from us. Yes, we have a long way to go, but we're starting to behave like a real company. Maybe we can turn this ship around and get back on track (that's a mixed metaphor, I know) before losing any more customers to our competitors.

April 11

Dear Diary,

While Noll spent most of the week calling customers, we anxiously waited for him to share the details. His report shocked everyone: Most of our customers never submitted firm orders for our products – it was just wishful thinking on Sal's part. Given the uncertainty based on the new standard, customers apparently were waiting to see who could back up their press releases with actual product.

That is not a bad strategy on their part. There's no risk for customers to wait on the industry to deliver products because adherence to the standard makes all competing products compatible. Such a standard – assuming it can be realistically met – effectively "commoditizes" a product at launch, giving customers huge leverage to purchase based on price. And our company was bringing up the rear in this race.

So for nearly a year we've been beaten up about deliver dates, when in fact many of the "commitments" that Sal held up to use weren't commitments at all. In fact, of all the orders we planned for, only one of the major customers had a critical need. I guess if no one else in the organization was trying to force us to move quicker, Sal's approach could be commended. But we needed more from him. There was no long-term planning with volume, sales price, channels, etc. – basically none of the 4Ps. On top of this, Sal frequently missed team meetings (busy lining up that next job?) where he should've been a permanent fixture. Sentiment for ol' Sal quickly went from "we'll miss you" to "good riddance." Remember that earlier term "tool"? Well I guess we can add another to the list.

But as project manager, I couldn't afford to dwell on these problems – my job is to quickly assess the situation, understand the root cause, correct the problem and move on. Sitting around whining just delays things further and consumes valuable resources. (Plus it hurts my throat.) As much as the news from Noll was alarming, we finally have known customer orders with due dates that are entered into our MRP system.

We also have a revenue stick to whack Dick with – the aforementioned "open orders" report that includes orders and more importantly, attendant revenue. At this point, we need to wait for the contract manufacturer to agree to our plan to hire lower-cost labor and then present the latest labor costs to Dick for his approval. I know the cost still won't be good enough, but it will be lower. And the plan gives us flexibility to address any fluctuations in demand.

April 19

Dear Diary,

A few of us were getting antsy, so we called the CEO of WBS a couple days ago to ask about the labor proposal. Except for some minor changes, he's in. Now we have everything in place for another fun-filled, esteem-building meeting with Dick. We can't accept another delay from him, since we're already significantly behind schedule for pre-production. Each of these meetings has been getting progressively less "warm n' fuzzy." I think we'll all be able to maintain our professionalism, but we can't defer to Dick's indecisiveness another day.

Meanwhile, there's one more item we forget to address – namely, how the hell are we going to test the product, since up until this time all testing was done by Design? Testing was part of my original schedule, but we never got around to it because of our focus on board design. Now it's in the "critical importance" column. We're still putting the cart before the horse – the product isn't actually working "to standard," so trying to figure out how to test it for pre-production is going to be another herding of the cats. All of these problems are on the low-band product. I hope they aren't repeated on the high-band product or I'll need to find me some powerful tranquilizers.

Some days it seems like our project is stuck in a traffic circle and no one knows which of the many roads that spin off the circle we should take. Fortunately, we have an experienced test engineer on our team, and he decided to travel to see our designers to understand what can be gained from ongoing design verification testing (DVT). Meanwhile, in order to focus the team on testing requirements, I quickly fashioned a simple schedule of major features we have to test and gave it to my counterpart prior to his trip. He'll modify the schedule based on information gathered during the trip so we'll have a decent start point for working through testing requirements.

I'm already dreading another meeting with Dick to get his approval for the equipment needed for testing. Like I said earlier, he's our biggest bottleneck and doesn't realize – or he won't acknowledge – this fact. The other important fact is that our managers lack the stones (balls) to challenge this situation – which exacerbates the problem.

In fairness, this is like déjà vu all over again. Many of the problems that plague this project I've seen throughout my career. If project managers are supposed to be on never-ending journeys of discovery, I suppose I'm just adding another stamp or two to my PM passport. One of the extraordinary lessons I've learned this time around: I never asked about the rules of the game before taking the job and joining in.

The problem with playing the project management game with my current employer is that Dick makes all the rules, changes them when he pleases and never lets anyone else in to improve the game. Project managers aren't dictators, and they're not secretaries either. But on a scale with dictator on the left and secretary on the right, somehow I've managed to get positioned to the far right. No, I don't want power for power's sake. I do want to be in a position to help drive revenue for my employer by satisfying customers. It's not ego – just the surest way I know to make sure I'll receive a regular paycheck.

April 25

Dear Diary,

In an attempt to exit the traffic circle, Operations decided we needed to meet with King Richard once more to see if we can get the approval to proceed with WBS on the pre-production build. This time we went in armed with potential revenue that's at stake if we're actually able to deliver functioning units under our renegotiated cost for labor. No surprise that Dick doesn't take long to attack our numbers on labor costs, even though there are firm revenue dollars that offset the cost of production.

He seems to occupy this personal financial dream world, where there's no need for any hard evidence to back up his numbers. Dick even made some off-hand comments about "your convoluted spreadsheets," the ones we worked up to try and address his convoluted questions about cost. These spreadsheets were set up to be able to recalculate labor costs based on his concerns. I mean, convoluted never looked, or performed, so good!

As the meeting wound down to a close, Dick got that all-too-familiar look of indecision on his face – his typical, "I'll get back to you in a couple days" expression. He started to say he wanted to give the cost numbers some more thought, and reflexively I rose up out of my seat (imaging my hands around his throat). No meeting etiquette points awarded there, but it did wipe that look off his face. It was time for Richard's indecision juggernaut to end.

I empathically stated that it would take weeks, maybe even months, to get any CM up to speed, and that the time to start is now. If we don't start this week, we might lose WBS altogether, even after weeks of negotiation and relationship-building. We'd end up delaying the program another month – a situation I find totally unacceptable. I was on a roll, so I also told Dick that moving to WBS now will free up our design engineers to transition to high-band product line, knowing that these words would be music to his ears. Dick looked at me, stared briefly at my boss – and said: "OK, go ahead with them. But make sure you keep track of costs." Game, set, match! We finally have a decision, and it didn't take weeks.

It's taken almost a year, but Operations is finally in a position to set up low-band production of early customer trial units. Of course, the design still isn't finalized, testing needs to be defined and developed, WBS needs to hire and train more contract labor, and customers expect critical units to be shipped next month. But at least the majority of the activities needed to drive to this point are within the jurisdiction of Operations.

We can now make a meaningful, measurable contribution to the organization after months of project delays. The truth is, as much as project managers and teams talk about making up time to offset delays, this is pretty much impossible. Unless you can attack a delay immediately, the lost time is usually permanent. But there should be fewer of them now.

I'm feeling better. So was my Operations team after I bought the first two rounds down at (the appropriately named) Murphy's Pub. Here's to project management done well.

Learning #11: Phenomena

Murphy's Law

For those of you hibernating for the last half-decade, Murphy's Law is a popular rule in Western culture that generally states: If something can go wrong, it will. More to the point, Murphy's Law posits that in any given situation, things will go wrong, in the worst possible way, if you give them a chance. Concerning the situation before us, it's possible that one member of the team is a descendent of Murphy (whoever and whenever he was) because of the seemingly endless string of problems encountered throughout.

Teams rarely factor in Murphy's Law. Instead they estimate tasks based on an "all success, all the time," best-case schedule, even though this scenario rarely ever happens – and we can go all the way back through history to demonstrate that. I'm guessing that even prehistoric men in prehistoric times faced project issues, like when they attempted to remodel their caves only to be done in by bats or rock slides or springs bubbling up. A predictable inability to ably predict a project schedule is something natural evolution has yet to mitigate.

If teams, and more importantly their corporate cultures, aren't prepared for things to go wrong, any slight problem is going to throw everyone off. Getting back on schedule will be nearly impossible. Imagine a car trip that, in a perfect world, should take four hours – 240 miles @ 60 mph = four hours. Assuming, of course, no traffic tie-ups, stop signs or red lights, bio breaks, flat tires, inclement weather, "slo-mo" O.J. police chases, children screaming "Are we there yet?" or inaccurate MapQuest directions.

What if, halfway through the journey, we experience a flat tire that takes 30 minutes to fix? Now we have to travel the remaining 120 miles in 1½ hours, which means traveling 80 mph to make up time – and going 80 mph is illegal. You get the point. And as more "something's" go wrong, there's just no way to stay within the original goal of four hours. Hence we're delayed. Our fabled team has been Murphy's Law'd to death, with Richard constantly stuck on the toilet pondering critical decisions.

What's the point of pointing this out? Programs – everyone's programs – get delayed all the time. It's effectively a law of nature, right? So what can really be done to help offset delays? Should we even be concerned since Murphy's Law "disadvantages" everyone?

Sure we should. Murphy's Law only seems like an unalterable law of nature. Set up a process at the beginning of a project to account for delays – all those "what if" scenarios I mentioned earlier – and you'll break the law, and gain a competitive advantage over all of those enterprises who resign themselves to the "inevitability" of Mr. Murphy. Form a team whose members have skills needed to quickly address "random" situations, whether caused by Murphy, a competitor, your own marketing team or government (for example, RoHS/WEEE directives).

What if, prior to our little road trip, we trained with a NASCAR pit crew and equipped the family SUV with quick-change tires that are standard for that sport? (Just work with me here.) Then, 15 seconds after the flat, we'd be back on the road – letting the air out of Murphy's tires and eyeing a new record of 14 seconds for our next flat. Any team's success is measured more on its ability to handle the unexpected than its ability to do the expected on time. After all, the team sets scheduled tasks and durations, but unexpected problems come with no set of instructions or an obvious fix.

In our favorite team's case, it certainly lacked the upfront planning that would have at least given it a fighting chance at success. Add to this all the problems encountered, both through its own actions (or inactions), as well as those of outside forces, and its failure to produce a sellable product in a timely manner was a given. This team apparently doesn't possess the organizational skills required to attack the problems as a team, leaving it up to individuals to function within their functions.

Successful teams look at all problems as a team issue first, with expertise coming from the appropriate function as needed. Many of the folks in our story, on the other hand, are looking to blame one particular organization – Operations – even when that group tries to keep its eye on the ball and doesn't join in to play the blame game.

No matter what happens during the life of a project, the project manager must never lose sight of the end goal – even if he or she has to create one. One other thing about project managers: They have to be like a major league shortstop that's just booted a ground ball. That is, they have to have short memories when it comes to the more nightmarish aspects of the game. Get it out of your mind and get on with it.

We've taken many wrong turns throughout our journey; however, we have finally arrived at the final destination, a place where we can start producing something our customers want. Operation's perseverance to guide the team with detailed schedules, action items, constant communication, use of best-class suppliers, and decision battles with Richard have come to fruition. Now that we've arrived, the journey really wasn't that bad.

Chapter 12

THE ANTIDOTE

The 4Ps of Project Management

I mentioned early on that my goal for this project – this book – is to help you appreciate the benefits of project management, prepare for inevitable problems, recognize factors that are consistent among projects, and keep a sunny outlook. And I want to revisit "The 4P's of Project Management." They are:

- Process – your project's "rules of the road"
- People – by far your greatest asset, and probably your greatest headache
- Parts – all resources (animate and inanimate) available to your project
- Phenomena – events, issues, attitudes and who knows what else that will work to propel or stall your project

These are the key elements that must be addressed if you're going to have any chance at managing a successful project. Don't confuse these 4Ps with "The 4 Ps of Marketing": product, price, place and promotion. Everyone approaches project management from their own particular perspective, influenced by their own personality and experience, and operating within the specific guidelines of their organization and its intrinsic culture. But those enterprises that address process, people and parts, and account for phenomena, give their project manager a running head start. It's still up to him or her, of course, to lead the team to that desirable end goal – a new product, service or achievement that enhances revenue, earnings or both for their company.

Process is the recipe for the delicious concoction; it includes all instructions and ingredients needed to create something that truly satisfies the customer's palette. *People* are required not only to make the dessert, but to handle and shepherd the ingredients, equipment and other items necessary to produce the successful result promised by the recipe. *Parts* are the actual ingredients, equipment and such. And *phenomena* are the "lumps in the mix" or "ghosts in the machine" that get in the way of following the recipe or operating the oven, and do harm to that to-die-for dish that everyone envisioned when they first entered the kitchen.

When you think of something that seems as commonplace as a dessert – say, a chocolate cake with chocolate chips, espresso beans and chocolate frosting – it's easy to forget that a lot of things can go wrong in the making. Ask any baker. Now replace dessert with a new, low-frequency product that's going to be sold worldwide and has to meet a plethora of regulatory and emission standards. Chocolate cake is "cake" by comparison.

Perhaps the most important key to success is this: Make time early on to thoroughly consider and address your project's processes, people, parts and phenomena before trying to accomplish anything. Why? Because it's too costly and time consuming to go back once the momentum of the project has started and expectations about end dates have been broadly communicated.

Process

Chapters 1, 2, 4 and 5 dealt with process issues, and the lessons they address are:

- Begin at the end
- Define a means to the end
- Specify the means
- The devil is in the details

People rarely do too much up-front planning. It's usually the opposite, and then, months later, we hear: "If only we thought of this earlier; we wouldn't be so far behind schedule, or over budget, or both." It happens when we plan a home improvement project, prepare for a family vacation, or go to the grocery store (without a shopping list).

Why do people pay professionals to build their homes or remodel their kitchens? Because professionals know the process, employ experienced people, understand how to procure parts and, most importantly, handle unexpected phenomena well. It doesn't matter how smart you are before heading off into "Project Land." If you aren't prepared, you're inviting disaster.

Good processes have been fine-tuned over time and cover those otherwise missing activities and initiatives that have made the difference between success and failure. It's the learning that comes with experience that makes us better able to manage things in the future. If you don't have any processes in the beginning and just sort of make things up as you go along, you're virtually guaranteed frustration and, from a business standpoint, a project that will be over budget, late and doomed to failure.

In our case here, the company's leadership stayed within its realm of expertise, expecting that the collective smarts of the group would result in success. Well, if success is defined as getting something working with the benefit of unlimited time and resources, our team won. But if there are limits on time, money and people, and if you're supposed to make a profit, they lost big. In the real world there's no such thing as unlimited anything, except potential problems.

People

Chapters 3, 6, 8 and 9 dealt with people issues, and the lessons they address are:

- Humor helps
- Old habits die hard (if at all)
- Leader
- Communications

Team members are like the opposite sex: You can't live with 'em and you can't live without 'em. Or maybe they're like an extended family where, based on statistics, there's going to be an Uncle Raffi – the relative no one wants to sit next to at the reunion picnic. We have a choice when it comes to courting and then marrying our spouses. Of course, that level of scrutiny isn't always possible when it comes to the composition of a project team. Project managers relish programs where the selection of some or all of the team is squarely in their hands. More often, there's little involvement by the project manager, as these decisions are left to others – usually department heads and upper managers – in the company.

Happily, processes can be the great equalizer, since they force the different characters on a team to get in step with the overall program. There's no ambiguity around expectations to perform based solely on an individual's preferences, as each member is expected to abide by the rules. Just look at a football team – the members play their positions based on established guidelines, deviation from, which can cause disastrous results.

Moreover, team members are expected to not only fulfill their role, but to help out others when needed. Here, a collective group with different skills and roles succeeds if it plays within the parameters of established processes. Processes also improve communications, enhance leadership and mitigate old habits. The only thing not accounted for is humor.

Humor is more than a necessary ingredient for projects – it's an essential ingredient. Without humor to occasionally break the tension and knit team members together, the frustrations that come with every project will fester and hurt the team's ability to manage the unimagined. Teams that have fun together perform better. They more efficiently remove roadblocks and are more in tune with each other's and their customers' needs. Humor energizes and helps team members look forward to going to work each day to tackle the next part of the project. That old adage, "Do something you love to do," applies here. Working with people who can laugh and joke together makes any project more loveable.

Parts

Chapter 7 dealt with parts issues ("Parts is parts."). You might think mine is a narrow concern for parts that fails to take into consideration all of the required inputs needed for a successful program. What about equipment, tooling, test fixtures, real estate, consumables, and all the other "hard assets" that go into developing a product or service. Well, you think right. As I use it, the term "parts" is a catch-all for everything that needs to be acquired, operated and handled in order to fulfill the project's goals. It's just that parts are such an obvious problem with our featured team (plus it starts with a "P") that it made sense to highlight the name.

The supply chain has become a very important aspect in managing today's businesses, as technology advances are creating new products and services almost in the blink of an eye. Customers demand the best, most feature-rich products as soon as possible and follow-on models to replace them in the near future. Getting a head start on parts, equipment and consumables that are inputs to a product allows a company to reduce time-to-market and establish its presence quickly to improve its profit.

It's just not feasible anymore (if it ever really was) for a team to wait until its product is perfect before revving up the supply chain. Doing so could result in parts for today's product arriving even as competitors are launching their next great mouse trap. Parts are a parallel project that must be managed with the utmost care if the corporation is going to succeed. Bottom line: In today's environment, customers won't wait for anything. Why frustrate them by taking your eye off the supply chain?

Phenomena

Chapters 10 and 11 dealt with phenomena, and the lessons they address are:

- Step back to move forward
- Murphy's Law

Let's be realistic: Even if a project is well planned, with a tested commercialization process, excellent people and a parallel blueprint for parts (or other items required by the project), phenomena – unexpected, unwanted developments – are going to occur. So why do project teams continue to fall into the "blue sky" trap, assuming that everything will go perfectly with their project, over and over again?

In my experience, humans don't want to admit that they're not smart enough to plan for the unplanned. Second, the initial enthusiasm of a new project clouds people's judgment. Third, we always want to sound positive for our bosses. And finally, no one likes change in the form of problems that get in the way of success. These common human traits will imprint onto the entire team, and then you get what's called negative momentum.

But how do you plan for unknown phenomena? Having experienced team members helps because they've faced phenomena and have lived to tell the story. Nothing is better than having team members who are "project hardened" and not intimidated by problems. These individuals can educate the team early about potential roadblocks they encountered on other projects and how they were overcome. It's up to the project manager to make sure this experience is included in the schedule. He or she might have to fight for it early on when team members are being selected.

Just adding time to the project schedule based on an expectation of some number of snafus, does little good (even though it does help set expectations in terms of the project's timetable). What really has to happen is to figure out a way to address an unexpected event before it can gain any traction. This requirement is met, at least in part, by a clear end goal, well-defined processes, efficiency around parts, and team experience. Last but not least, an important key to mitigating phenomena is attitude.

There has to be certain cockiness or swagger within the team that dares Murphy to mess with its progress. The team takes the Old West Sheriff mentality: No one messes with my town on

121

my watch. These types of teams rise to the occasion and quickly replace their functional hats with trouble-shooting helmets and attack the problem with precision.

How do you create such a team? That could be the topic of a whole other book. I've been part of experienced and inexperienced teams that possessed a never-say-die attitude. The only common thread between them was a clear goal and an espris de corps that led to a determination not to fail in its pursuit of that goal.

And then, sometimes, it comes down to positive phenomena. As much as project managers plan out every conceivable aspect of their work, good old-fashioned dumb luck can sometimes make or break the program. Of course, some people say you're your own luck. In any event, these are the times when the whole in terms of team chemistry, is better than the sum of its parts.

I hope you get the opportunity to be part of a team like that. I hope you get to lead one as well.

Chapter 13

Final Entry

Stardate log May 1st

Dear Diary,

So what did Operations do to keep the project moving toward production? As time passed, we adopted sound processes and learned from our experiences. The idea that we should "start at the end" in terms of agreeing on product features, structure, volume and cost became much clearer. It enabled the project team to finally start reading from the same sheet of music.

In addition, our ability to "define a means to the end" regarding key milestones improved as Design got closer to creating a functional product. This allowed Operations to quietly take ownership of the project. Why? Because there was no way we could prepare the factory without Design, Marketing and Service completing critical tasks that were inputs into Operations. This enabled us to start producing saleable products.

We documented our activities as well as those from other departments that hadn't been doing as good a job recognizing tasks they needed to complete in order to launch the product. This started out as a way to ensure that Operations could speak confidently to due dates with Richard – something he expected of us because this discipline was already an intrinsic part of our organization.

As time progressed and the day-to-day chaos escalated, we scheduled hourly meetings every day and invited representatives from Design, Marketing and Service. The meetings enabled us to "specify the means" and thereby, take over the project leadership without anyone realizing we had done so. Everybody was so concerned about sharing bad news with Dick, they left this job up to us. With nothing really to lose and everything to gain, we drove the schedule.

These meetings, painful and grueling as they sometimes were, made it clear to everyone on the project team what needed to be done. They forced people to communicate – to "talk the talk" – and in ways brought the team closer together, as if we were in the same room. There was no hiding behind functional walls because we were all part of a process that required each of us to successfully launch the product.

The discipline that was lacking from our head of design was now in place – in the form of my project schedule and a handful of spreadsheets that spelled out everyone's role in achieving customer dates. Each meeting allowed Operations to take more control over the project, because no one wanted to be the reason for missing a critical milestone. We finally started to look outside the organization and take into consideration customer expectations.

Even though we should've begun many of these activities much earlier, it's almost never too late to start. We found that just as bad things can snow ball, it's also true that one good thing can lead to another. Initial successes helped minimize functional fighting and focus everyone on activities required to move program forward. Documentation doesn't have to be a five-hundred-line Microsoft Project schedule – just a simple spreadsheet that includes pivotal tasks that need to be addressed.

So, to borrow a phrase from my children, "Are we there yet?" No, we still have a long way to go before driving any significant revenue. But processes are getting defined, people required to launch the product are in place, parts are arriving daily and phenomena are being handled with greater diligence. Things are definitely starting to look up.

If we're honest with ourselves, we'll admit that we're all responsible for the project delays and problems. They affect our customers; the individuals who ultimately pay our salaries. The Operations team could've and should've done more than it did to improve time-to-market. After all, many of us were part of successful projects in the past and we knew what it took to be successful. But people have a hard time being the "boat rocker" or the "whistle blower," so the team morphs into whatever shape the corporate culture allows and continues down its particular path.

Despite all that, what helped us were ingrained habits, born of experience, to document activities so we could understand the probability of meeting dates and launching product. Who would have guessed that we could've brought others into the "document tasks" fold to help the project make some discernable progress (and maybe ultimately succeed)?

Most of my frustration was due to the lack of a clear road map to the final destination – a leading-edge product that would delight the customer and make the company significant revenue. It was up to me to map out my portion of the trip and to cooperatively include others in the journey. Eventually we got a map everyone could understand – one that was going to help everyone complete the journey.

The map allowed individual functions to find better ways to make the trip a success. What we should've done in the beginning was develop our own AAA TripTik, with all of the distances, construction sites, roadblocks and detours already marked. Then we could've spent an extensive amount of time planning the project based on this TripTik before we ever got in the car. Instead, we all jumped into the car and started driving with only the vaguest destination in mind.

You live, you learn. I've learned – or more accurately, I've been reminded – how much I enjoy this work, and how essential it is to business and personal success.

- END -

Character List:

Supplier	ACME Inc.
CM	We Build Stuff
Dean Sign	Design Manager
Will Invent	Lead Engineer
Isai Goodenough	Engineer
Cody Smith	Software
Peter Parts	Supply Chain
Manny Factura	VP of Operations
Sal Moore	Sales/Marketing
Noll Itall	Sales Representative
Richard Cranium	Business Unit (BU) Director
Sarbenes Oxley	COO or Finance

Chapter 14   References

Crawford, Merle, Di Benedetto, Anthony, *New Products Management*, McGraw-Hill, 2006

Clark, Kim B., Wheelwright, Steven C., *Managing New Product and Process Development*: Text and Cases, The Free Press, 1993

Tabrizi, Behnam, Walleigh, Rick, *Defining Next-Generation Products: An Inside Look*, Harvard Business Review, November – December 1997

Smith, Edward, Wheelwright, Steven C., *The New Product Development Imperative*, Harvard Business Case, February 1, 2001

Iapoce, Michael, *A Funny Thing Happened on the Way to the Boardroom: Using Humor in Business Speaking*, Wiley, 1998

Alpern, Lynne, Blumenfeld, Esther, Humor at Work: *The Guaranteed, Bottom-Line, Low Cost, High Efficient Guide to Success Through Humor*, Peachtree Publishers, 1993

Fahlman, Clyde, *Laughing Nine to Five: The Quest for Humor in the Workplace*, Steelhead Press, 1997

LaCroix, Darren, Segel, Rick, *Laugh & Get Rich: How to Profit from Humor in Any Business*, 2000

Bowen, H. Kent, *Project Management Manual*, Product Number: 9-697-034, Harvard Business School Publishing, 2002

Tuckman, Bruce W., *Developmental sequence in small groups*. Psychological Bulletin, 63, 384-399, 1965.

About the Author

Donald Pillittere is an accomplished product manager with proven success in developing and implementing long-term strategic vision to achieve profitability. His leadership has enhanced profit performance in a variety of disciplines both by increasing sales and significantly reducing expenses, without sacrificing quality or customer satisfaction. These results have frequently been achieved while simultaneously beating time-to-market deadlines. This results-driven professional's 30-year career has encompassed digital imaging / film scanning; medical imaging and health sciences; and process controls for both light manufacturing and heavy industrial applications.

Currently Manager of Portfolio Strategy and Supplier Quality at Exelis Inc. (formerly ITT), Mr. Pillittere manages supplier quality for highly advanced government subsystems. Prior to working at Exelis Inc., he was Director of Product Engineering and Program Management at Transonic Systems Inc., where he managed the development and introduction of numerous medical flow measurement products for both clinical and research applications. Previously as Worldwide Product Manager for the Eastman Kodak Company, he launched numerous award-winning scanners that significantly outperformed sales projections and exceeded profitability goals. He also has a successful track record in the international arena, with experience in Asia, Europe, Latin America, and Japan, as well as North America. A sought-after expert, Mr. Pillittere was frequently quoted in corporate press releases, as well as various periodicals and professional journals.

Sharing his wisdom and real-world experience as Adjunct Professor at his alma mater, Mr. Pillittere has taught graduate courses in Operations, Supply Chain and Global Facilities Management, Manufacturing Strategy & Tactics, and Managing Manufacturing Resources at the Rochester Institute of Technology's College of Business since 1999. He successfully combines his industrial experiences with the theory and principles of operations management to offer students practical, proven tools and approaches that can be applied in a wide variety of business settings. A gifted public speaker, Mr. Pillittere creates a dynamic, challenging, and fun learning environment that he adapts to the specific needs of each class.

He earned a Bachelor of Science degree at the State University of New York at Buffalo in Electrical Engineering; and holds a Master of Business Administration degree from RIT's Executive MBA Program, from which he graduated With Honors, as a member of Phi Kappa Phi and Beta Gamma Sigma Honor Societies. Mr. Pillittere is also the author of a second book *Hot Chili and Cold Beer for The Project Manager's Soul* as well as 4 business cases available on Harvard Business Review and Ivey Publishing.

Clearly a highly motivated individual in his personal life as well, Mr. Pillittere is an accomplished marathon runner, having competed in events in New York City and Chicago. Happily married for 30 years, he and his wife are the proud parents of three children, with whom he resides in Spencerport, New York.

Reviews for Are We There Yet, Diary of a Project Manager

5 out of 5 stars – All too familiar episodes

By Lino Nobrega, March 28, 2010

"I found myself smirking while reading this book, it takes you on a ride through Corporate events, that are probably common in most work places. Well detailed, and I recommend that you learn from Don, he knows what he's talking about. Buy the book for your entire team, not only will they appreciate it, they will get great insight and hopefully avoid some of the common mistakes."

5 out of 5 stars – A book for project managers and team members stuck working in the real world

By Gene R., April 10, 2011

"Having read numerous books on project management methodology, technique and practice I found Don's book to be practical and realistic in regard to what ACTUALLY happens in the REAL WORLD of project management. Don't get me wrong, as a certified PMP I think the PMBOK (Project Management Book of Knowledge) is a great resource describing how projects should go. Don simplifies the process into what he calls the four "P" of project management People, Parts, Processes and Phenomena. His real world stories will send the message home and will even make you laugh, especially if you have spent much time in corporate America. It is an easy read with great content; I recommend the book to anyone working in project management or as a contributor on project teams."

PM Forum

Published in *PM World Today* – August 2010 (Vol XII, Issue VIII)

PM World Today is a free monthly eJournal - Subscriptions available at http://www.pmworldtoday.net

Excerpts from review

Introduction to the Book

This book focuses on all elements of the project management lifecycle, utilizing real-life examples in a diary format that is easy to relate to and understand.

The project management 4 P's (Process, People, Parts, and Phenomena) enables the reader to become more knowledgeable by handling obstacles (internal and external), thereby providing a foundation to be better equipped to succeed. Being that each project is unique in nature, this book focuses on important aspects such as working culture, internal processes, and the team environment. Ultimately, project characteristics that is consistent among all projects.

Overview of Book's Structure

The diary format symbolizes a day-to-day chronicle of a project team developing a series of new products. This stimulating read has short chapters, providing an engaging look at the trials and tribulations of project management; accompanied by many 'aha ha' moments in this vastly dynamic profession. Each chapter is structured where the project obstacle is stated upfront. The chapter discussion highlights the current scenario the team is facing, and the chapter ends with what the team could have done better to mitigate the difficulty.

Highlights: What I liked!

The flow of the book is a story in its entirety, based on real-life experiences and is very easy to comprehend. The comical highlights add humor to stressful situations.

"One needs to see the light at the end of the tunnel...and sometimes it's an oncoming train!" Additionally, the 'character list' was well thought out, creatively written, and a 'nice to have' reference. Furthermore, the final chapters tie the entire book together, and summarize the four 4's of project management (Process, People, Parts, and Phenomena), relating examples to applicable chapters.

Hot Chili and Cold Beer for the Project Manager's Soul

Donald A. Pillittere

4Ps of Project Management Series

Project manager are a different breed, hardened through years of missed milestones, budget overruns, dysfunctional teams and unrealistic management edicts with the battle scars to prove it. Besides the CEO, no other person has a greater impact on the profitability of a corporation than a project manager. This indubitable book provides real-life experiences of project managers in the quest to succeed against all odds. Presented as a series of short stories each chapter focuses on how the project manager turns "Chaos into Cash" using the four "Ps" of the project management: Processes, People, Parts and Phenomena.

So relax, pour a cold beer, grab a bow l of hot chili and enjoy The Good, the Bad and the Utterly Random concerning one of the most essential, unpredictable and unsung responsibilities in business.

Available in hardcopy and ebook

The Project Code

Heralded as one of the greatest inventors of all time, DaVinci would have failed as a program manager. It took hundreds of years for most of his designs to be built and successfully tested. DaVinci had The DaVinci Code and what he needed was The Project Code, the four P's of Project Management: People, Process, Parts and Phenomena in quantifying and eliminating risk. This book provides a practical application of The Project Code so that project managers can control risk and navigate the project to a successful completion.

So get in your Snuggie, grab a bowl of popcorn, and enjoy The Good, the Bad and the Utterly Random concerning one of the most essential, unpredictable and unsung responsibilities in business – Project Management.

CPSIA information can be obtained
at www.ICGtesting.com
Printed in the USA
LVOW13s0333180117
521347LV00006B/244/P